The Lonely Sentinel

The Lonely Sentinel

Fort Clark:
On Texas's Western Frontier

Caleb Pirtle III & Michael F. Cusack

EAKIN PRESS
Austin, Texas

HOUSTON PUBLIC LIBRARY

Friends of the
Houston Public Library

R0127702518
txr

FIRST EDITION

Copyright © 1985
By Michael F. Cusack

Published in the United States of America
By Eakin Press, P.O. Box 23066, Austin, Texas 78735

ALL RIGHTS RESERVED

ISBN 0-89015-508-9

For the memory of
John Whitten

"After all," he said, "we're only passing through . . ." The old man then gave a big smile, twisted his handlebar mustache, and had a sip of Madeira — vintage 1849.

Letter to Mary Douglas (Read) Brickhouse, 1877

Fort Clark is desolate, and it is deathly hot. I am twenty-three and a West Pointer and caught in hell — a continuing war. My soul is in anguish over the plight of some of the good settlers — and in like anguish for some of the Indians. (Anguish — and my God, Sister, I am a soldier.)

In a hundred years we'll all be gone . . .

And only Fort Clark will still be here.

— Lieutenant Robert Doddridge Read
Assigned with Cavalry to Fort Clark
After 1877 graduation from West Point

Contents

Foreword	vii
Preface	ix
1 A RAGE OF FEAR	1

A defiant civilization of Indians battled the surge of white settlers who pushed their way into the unclaimed territory of Western Texas. The Indians would fight, even die, before they let the lands slip forever from their determined, yet fragile, grasp.

2 THE GUARDIAN OF THE PRAIRIES	10

By 1852, the U.S. Army had decided to establish a chain of forts across the western fringe of Texas to bring a measure of protection to a harassed country. One post, nailed to a rocky ridge of limestone that curved along Las Moras Creek, would be known as Fort Clark.

3 SURVIVING IN A SELFISH LAND	25

The land had no friends. It was hard and sometimes deadly, and it had no tolerance at all for those who came to stay. The soldiers of Fort Clark viewed themselves as condemned men who had been banished to a condemned parcel of earth. They patrolled a vast prairie, searching for Indians who struck quickly, then hid away in the mountain sanctuary of Mexico.

4 THE CHANGING OF THE FLAGS	47

When the War Between the States spilled into Texas, the Union troops marched away from Fort Clark, leaving it in Confederate hands. The rebels had little time to worry about the duel between North and South, however. The Cavalry at Clark was too busy trying to prevent Indians and Mexican bandits from conducting a "civil and fractricidal" war upon the barren Trans-Pecos.

5 INVASION IN A FORBIDDEN LAND 67

The War Between the States had ended. But West Texas lay at the mercy of renegades who found protection in a foreign land beyond the Rio Grande. Then Colonel Ranald Mackenzie was handed his unofficial orders to launch a campaign of "annihilation, obliteration, and complete destruction" of the tribes, even if it meant invading Mexico. And it did.

6 THE LONG MARCH NORTH 92

After his successful raid into Mexico, Colonel Mackenzie and his Fourth Cavalry trekked northward, marching from Fort Clark to the Staked Plains and the day of bloodshed that awaited them in the rugged abyss of Palo Duro Canyon.

7 THE PATH OF THE WHIRLWIND 107

The Indians, at last, were on the run. And the cavalry from Fort Clark gave them no peace and no quarter as war spread westward. Renegades could no longer escape the tracking ability of the Seminole Negro scouts, and they could not outfight these refugees from the Florida swamplands. The Comanches fled the wrath of Lieutenant John Bullis. And they called him "The Whirlwind."

8 BATTLE OF THE BLOODY BEND 121

In Mexico, a revolution erupted in 1911, and the soldiers of Pancho Villa fanned out along both sides of the Rio Grande in search of food, supplies, and ammunition. They looted, and they killed anyone who stood in their way. The cavalry of Fort Clark would have to stop them, if they could be stopped at all.

9 THE COMING OF AGE 139

Although it became the stepchild of the U.S. Army, Fort Clark would send men to fight in two World Wars. And from its West Texas training ground came two of the military's most honored generals: Jonathan M. Wainwright and George S. Patton.

10 EPILOGUE 163

In 1945, the sentry of the plains was officially declared as surplus. A year later, the troops marched away for good. Fort Clark became an old soldier that had died, but has never faded away.

Photographs 173
Bibliography 215
Index 221

Foreword and Acknowledgments

Toward the end of my research on this book, I was sitting in a hotel room in New York City. Going over some papers, I came to the conclusion that I was running out of places to search for materials. I was sure that this was wrong. How could I have finished researching . . . when I just knew there had to be miles and miles of still unfound stories and pictures.

I flipped on the television set and a talk show was in progress. The host was interviewing the wife of a writer who had just died. It seems he spent twenty-six years researching a book on Japan and World War II. His wife kept urging him to stop the research and publish the work. "NO" he said, "I'm not finished." He never saw his book published.

Both the talk show host and the writer's wife came to the conclusion that there was enough material gathered to publish fifteen books. It took four years to edit the material into one book . . . a prime case of over-research.

I don't think any researcher likes to quit looking for material. The thrill of discovery never stops.

"Trite but True" . . . it takes a lot of people to put together a book project.

I would like to thank the following . . . University of Oklahoma-Western History Collection, Institute of Texan Cultures–San Antonio, Texas; Ralph Elder, Barker Research Center–Austin Texas; Daughters of the Republic of Texas Library–San Antonio, Texas; Dr. Peter Olch, Bethesda, Maryland; The Pentagon–Photography Depository, Washington, D.C.; Sergeant First Class Clyde Waltman–1st Cavalry Division, Fort Hood, Texas; J. A. Small's *True West Magazine*–Austin, Texas; Delores Raney–County District Clerk, Brackettville, Texas; George Wyrick, Brackettville, Texas; William Hilderbrand, Gainsville, Florida. Joe Elicson, Blanco,

Texas; Dr. William Miehr, Jr., San Angelo, Texas; West Texas Museum Association, Lubbock, Texas; Library of Congress, Washington, D.C.; National Library of Medicine, Bethesda, Maryland; National Record Center–Photo Division, Suitland, Maryland. San Antonio Public Library, Houston Public Library, Brown and Root, Houston, Texas; Jack McGerk, Smithsonian Institution–Washington, D.C.; Mrs. Agnes Vondy, Brackettville, Texas; Nat Mendelsohn, Fort Clark Springs, Brackettville, Texas; Mrs. May Ellen MacNamara, The University of Texas Humanities Research Center–Photography Collection. Joe Thomas, National Archives, Washington, D.C.; Bob and Cathy Conrey–Fort Clark Historical Society, Brackettville, Texas; George and Shirley Doak, Aledo, Texas.

Thanks to John Lindy, who hand-printed photos from negatives of old pictures, and Jim Hughes from the Deep River Armory in Houston was very helpful.

A very special thanks to Happy Shahan and his family, Brackettville, Texas, who first gave me the opportunity to see Fort Clark, and B. J. Burns, and Publisher Ed Eakin, Eakin Press, Austin, Texas . . . and Editor Shirley D. Ratisseau for her historical and editing expertise . . . and to Caleb Pirtle who put it all together from boxes of research material.

MICHAEL F. CUSACK

Preface

For years, on my journeys westward across the Trans-Pecos of Texas, I hardly glanced at Fort Clark. It was nothing more than a scattered collection of stone buildings — some restored, some crumbling — mostly forgotten or at least ignored. It gave the appearance of a ghost town, out of time if not out of place: silent, deserted, trapped in the heat and the dust of a harsh country.

The fort had its secrets, I knew, but the old outpost on Las Moras Creek kept them all to itself. Few ever talked much about the lonely sentinel. Not many beyond the outskirts of Brackettville either knew or cared about the role Fort Clark and its soldiers played in bringing peace to the prairies of Texas.

Other outposts whose guns pointed west had earned a certain measure of fame, or maybe it was just notoriety.

Some were praised.

Some were damned.

Most had become legends.

Fort Clark was different. It was an enigma, sitting there beneath the oaks and mesquites just south of U. S. Highway 90. Few were aware of just how old it was, why it had lived, or when it had died as an Army post. The memories of its battles had been blown away like dust across the plains. Its glories had faded. Its secrets had been buried in too many unmarked graves and in too many files that no one bothered to read anymore.

As Chief of Media Relations for the Texas Tourist Development Agency, I had the opportunity to work with Governor John Connally's task force that established the Ten Texas Travel Trails, guides to the scenic, recreational, and historic points of the state. In mapping out the Pecos Trail, about all we had to say about Fort Clark was that it "was one of the most historic forts in the Southwest. Established in 1852, it served until deactivated in 1946. Over the years many infantry regiments and most of the U. S. Army's cavalry units saw duty here. Now Fort Clark is a retirement and leisure home development."

It wasn't so deserted anymore.

But the secrets remained intact.

During my first year as Travel Editor of *Southern Living Magazine*, we published an article on the Fort Clark guest ranch as a hideaway vacation spot that offered both peace and relaxation. The story mentioned that "Fort Clark, as it was named, housed troops to protect the huge wagon trains made up in San Antonio and bound for California — always tempting targets for hungry, marauding Kickapoo and Comanche raiders. From antebellum days through World War II, this old frontier post was home to some of America's most famous fighting generals." However, in 1969, we believed that travelers were much more interested in the "Western posh" decor of the guest ranch quarters, as well as the chance to play golf, croquet, volleyball, and horseshoes upon the same ground where those fighting generals had walked.

The fort was still jealously guarding its secrets.

That, I would learn, was an injustice, one that needed to be rectified. Fort Clark had stood too long, seen too much, suffered too many times. It was haunted by such familiar, historic names as Ranald Mackenize, John Bullis, Phil Sheridan, William T. Sherman, George S. Patton, and Jonathan Wainwright. Its soldiers had chased Indians on the prairies and in the mountains, Pancho Villa into Mexico, and the Germans of two world wars across Europe. Its past was decorated with bravery and heroism. Fort Clark was simply too important to be either forgotten or ignored.

The lonely sentinel still has its secrets.

But, at last, a few of them have been told.

CALEB PIRTLE, III

1

A Rage of Fear

The empty land that was Texas had been tarnished by anger and tainted by blood that ran deep into the rivers of the forgotten. The Indians owned the soil, or so they said. To them it was sacred, the giver of life, and no one should take it from the calloused hands of their people. They stared with hard, solemn eyes at the dust of the white man's civilization as it crept slowly like a plague across the prairies where the buffalo roamed as free as the wind in the tall, dry grasses. Their world was shrinking, and they would fight, even die, before it slipped forever from their determined, yet fragile, grasp.

On the banks of the Oak Creek, the Comanche Chieftain Yellow Wolf turned his back on gifts that had been sent his people by the government. With hatred in his eyes he snapped, "It would be impossible to make white men out of Indians. You might as well try to make a dog out of a wolf."

He rode away.

And Yellow Wolf fell, driving stolen horses toward the far western mountains that hid the scattered remnants of his tribe.

A Texas Ranger stood atop the caprock and talked with a settler as the sun reached down and blistered the ground around them. He wiped the sweat and grime from a leathered face and gazed beyond the ragged ravines where the Comanche trail had ended.

"Old Yellow Wolf was kilt for his beliefs," the settler said.

The Ranger shrugged. "He had the belief that all the horses in the country was his'n," he answered.

The soil that rolled on toward the setting of the sun held the settlers' hopes and their dreams. It would also hold the graves of their children. The vast frontier beckoned with a promise it could not keep. The nesters came, even though they had been warned: *The best side of the land's up. For God's sake, don't plow it under.* And some farmers swore that they had tried to nail down a home upon plains where it was one hundred miles to water, twenty miles to wood, and only six inches to hell. Still the wagons kept rolling westward. The soil was dry, barren, and cursed with misery. Only sweat and blood —never rain — moistened it. But it was cheap. All that it would ever cost them was life, and it cost them plenty.

Men had more faith than vision. Hardships were more common than money. Pioneers didn't run from them. They endured them and fully believed that each sunrise brought them one day closer to peace on earth than they had ever been before. As William B. DeWees wrote a friend in Kentucky in 1839:

> The future is ripe with the most sanguine anticipations of prosperity, wealth, and happiness. Energetic and active emigrants are rapidly flocking into the country. We hope soon to be out of danger from the Indians, being protected by the military road, and with the rest of the world we are at peace. True, Mexico has not yet acknowledged our independence, but we have little to fear from her even should she again invade our country.
>
> I would still advise you to come to Texas. To all who would carve out for themselves fortunes at the expense of a temporary sacrifice of the luxuries of life, in preference to fretting away their lives in earning a subsistence in a country already filled with hungry competitors, I would say come to Texas; buy you farms, or in some way use your talents industriously, and be independent. Should the savages attack your homes, or even should our former enemies the Mexicans, again invade our peaceful delightful land, we have the satisfaction of knowing that those homes are richly worth defending.

Wagon wheels cut ruts in the caliche soil that became roads and small wooden cabins rose up above the shinnery. The buffalo were chased on further westward, and saddened Indians followed in the agony of their weary tracks. Katemcy, a Comanche chief, would say to the white man who trespassed upon the land of his fathers:

> Over this vast country, where for centuries our ancestors roamed in undisputed possession, free and happy, what have we

left? The game, our main dependence, is killed and driven off, and we are forced into the most sterile portions of it to starve. We see nothing but extermination left us, and we await the result with stolid indifference. Give us a country we can call our own, where we may bury our people in quiet.

There would be no quiet for anyone. Only solitude in a land so vast that no one could hear the cries of the frightened or the curses of the damned. The Indians came with the sudden swiftness of a summer storm, and they stalked the shadows of the night. They left behind cabins that crumbled and burned with the flames of their vengeance. They sought to cleanse the soil that the gods had given them, wipe away the ugly stain of those who dared stray into a land where they did not belong.

Texas, during the early days of the nineteenth century, shuddered at the sight of its own blood.

Josiah Wilbarger went down with an arrow through the calf of his leg, and as he tried to escape the ambush, a rifle ball slammed into his neck, again knocking him to the ground, paralyzed and unable to speak. The Indians stripped away his clothes, then tore the scalp from his skull. Wilbarger always swore that he felt no pain, but the sound of his skin being ripped away reminded him of distant thunder. At last left alone, he crawled to a creek, ate snails as he listened to the maggots at work inside his wounds, and waited for someone to ride his way and save him, and someone finally did.

At the Goacher homestead, Indians slipped upon two children playing near the edge of the woodlands. The youngsters, they decided, could not be trusted. They might cry out and warn those who had stayed inside the cabin. A steel spear was rammed into the little boy, killing him instantly. The young girl was taken prisoner, then slain as the savages stormed the cabin, leaving only death to weep behind the walls when they had gone. A man, his wife, a son, a son-in-law, and two grandchildren lay in silence upon the dark and bloodied ground. The daughter and her two-month-old child were dragged away. The night wore on. The babe cried with hunger. An Indian, weary of the wailing, ripped the infant from its mother's grasp and threw it into a lake to drown and be quieted at last. A screaming mother waded into the water and pulled her child from the lily pads. When another brave tried to slash the baby's throat, she grabbed a stick and knocked him to the muddy shoreline. She braced herself, waiting to die. But the Indians only laughed.

"Squaw too much brave," one said. "Damn you. Take your papoose and carry it yourself — we will not do it." He gently handed her the babe and walked away.

Months later, mother and child were bartered off at Coffee's trading house on the Red River for four hundred yards of calico, a large number of blankets, and a few beads.

Another mother, in another part of Texas, would not be so fortunate. Rachel Plummer was taken prisoner by the Comanches and Kiowas, and her agony was penned in the small diary she kept:

> My child was six months old, when my master thinking, I suppose, that it interfered too much with my work, determined to put it out of the way. One cold morning, five or six Indians came where I was suckling my babe. As soon as they came I felt sick at heart, for my fears were aroused for the safety of my child. I felt my whole frame convulsed with sudden dread. My fears were not ill grounded. One of the Indians caught my child by the throat and strangled it until to all appearances it was dead. I exerted all my feeble strength to save my child, but the other Indians held me fast. The Indian who had strangled the child then threw it up in the air repeatedly and let it fall upon the frozen ground until life seemed to be extinct. They then gave it back to me. I had been weeping incessantly whilst they were murdering my child, but now my grief was so great that the fountain of my tears was dried up. As I gazed on the bruised cheeks of my darling infant, I discovered some symptoms of returning life. I hoped that if it could be resuscitated they would allow me to keep it. I washed the blood from its face, and after a time it began to breathe again. But a more heartrending scene ensued. As soon as the Indians ascertained that the child was still alive, they tore it from my arms, and knocked me down. They tied a plaited rope around its neck and threw it into a bunch of prickly pears and then pulled it backwards and forwards until its tender flesh was literally torn from its body. One of the Indians who was mounted on a horse then tied the end of the rope to his saddle and galloped around in a circle until my little innocent was not only dead, but torn to pieces. One of them untied the rope and threw the remains of the child into my lap, and I dug a hole in the earth and buried them.

No one was safe as the Indians fought desperately to hold on to their sacred gift from the gods: the land.

War turned civilized men into animals, and they swore, in Biblical terms, to take an eye for an eye, and often they were as vicious, as cruel as the Indian raiders they damned. When the soldiers finally arrived, there were those who, without remorse, destroyed villages, raped the tribal women who were left behind, and turned children loose to run naked and crying across the plains. Indian scalps hung bleeding from their belts. And some troopers tortured their captives, peeling away their skin and burning them as grim reminders to other renegades who dared fight the U. S. Army. There were no saints, only frightened, angry men who would kill or be killed. It was a bitter war.

On the banks of Waller Creek, just outside of Austin, a raiding party of twenty-five, maybe thirty, braves rode down hard on Captain Coleman and William Bell as they trudged home through the cold January afternoon. In a small field, Bell lay dying, and his scalp soon hung from an Indian's belt. Coleman was captured and stripped of his clothes, then prodded along in shame and humiliation by the sharp points of the raiding party's spears. As they all passed Nolan Luckett's house, two small boys came driving cattle across the open field. One was taken prisoner. The other ran screaming toward the house. An arrow in his small, frail back silenced the screams as a winter's dusk draped itself around the landscape.

James Alexander pulled his old, cranky wagon to a halt at the head of Pin Oak Creek on the way to Bastrop. He and his sixteen-year-old son eased themselves down to the ground and began preparing lunch. It was a meal they would never finish. Indians suddenly rose up out of a ravine, so close to Alexander that their gunfire actually scorched his clothes with powder. The three Caddos and two Cherokees murdered both man and boy, horribly mutilated their bodies, looted the wagon, and killed the oxen before riding triumphantly away. Alexander had come to find peace upon the soil of Texas. There was none. His peace was eternal.

Texas sought to negotiate an end to the hostilities. Some of the Indian chieftains listened. Others turned a stubborn, deaf ear.

One Comanche leader, Pah-hah-yuco, told the Indian agents at Torrey's Trading House on the Brazos:

> Never give up your efforts to make peace with your Red Brothers. Whenever any of them come to see you, smoke the pipe of peace with them and give them good talk before they leave . . . And you must give them presents when they come. That will not hurt you, but if they should cut your meat off, that would hurt you . . . You that are listening to me think I am telling lies, but the Great Spirit who looks upon me now knows that I speak the truth.

Pah-hah-yuco did. He even returned a little white boy captive just to prove his sincerity. It wasn't a sign of weakness. Pah-hah-yuco had only spent too many years watching the wagon wheels cut deep ruts across his hunting grounds. The flood of humanity was coming, and it could not be stopped. His people could hold onto their lands for a while, perhaps, but certainly not forever. To survive, they would all have to learn to live with the white men and as the white man lived.

It grieved Pah-hah-yuco. But he could see no other choice. Other chieftains did. They were angry, and they would fight, and they would make the white man pay for his transgressions with the blood of his family.

Most were filled with hatred.

Death, for them, was a willing sacrifice to keep from losing the soil which held the ashes of their fathers.

For too many years, the Indian tribes had roamed as free as the winds that swept before them across the Texas prairielands. They wanted to be left alone. They wanted the white man to turn and go away and leave the plains as empty as they were before the settlers had come. All they could hear was the stampede of civilization drawing closer and drawing their freedom to an end.

One Comanche Chief, Mukewarrah, told the old Ranger, Noah Smithwick:

> We have set up our lodges in these groves and swung our children from these boughs from time immemorial. When game beats away from us we pull down our lodges and move away, leaving no trace to frighten it, and in a little while it comes back. But the white man comes and cuts down the trees, building houses and fences, and the buffalo gets frightened and leaves and never comes back, and the Indians are left to starve, or, if we follow the game, we trespass on the hunting ground of other tribes and war ensues.
>
> No, the Indians were not made to work. If they build houses and try to live like white men they will all die. If the white man would draw a line defining their claims and keep on their side of it the red man would not molest them . . .

But no line was ever drawn.

There were no fences to keep the settlers back. So they surged westward where land was cheap, yet so valuable to them. All it took to become a landowner were grit and guts and a rifle that was always primed. Men buried their roots deep and their graves shallow.

Ten Bears, the Comanche Chief, would one day stand before the horse soldiers at the great Medicine Lodge Council, and in a voice as cold as winter itself, point the final accusing finger for the death and destruction that spread across Texas like a disease that could not be cured. He said:

> My people have never first drawn a bow or fired a gun against the whites. There has been trouble on the line between us, and my young men have danced the war dance. But it was not

begun by us. It was you who sent out the first soldiers and we who sent out the second. Two years ago, I came upon this road following the buffalo, that my wives and children might have their cheeks plump and their bodies warm. But the soldiers fired on us, and since that time there has been a noise like that of a thunderstorm, and we have not known which way to go . . . Nor have we been made to cry alone. The blue-dressed soldiers and Utes came from out of the night when it was dark and still, and for the campfires they lit our lodges. Instead of hunting game they killed my braves, and the warriors of the tribe cut short their hair for the dead.

So it was in Texas. They made sorrow come in our camps and we went out like the buffalo bulls when the cows are attacked. When we found them we killed them, and their scalps hang in our lodges. The Comanches are not weak and blind, like the pups of a dog when seven weeks old. They are strong and farsighted like grown horses . . .

If the Texans had kept out of my country, there might have been peace. But that which you now say we must live in, is too small. The Texans have taken away the places where the grass grew the thickest and the timber was the best . . . The whites have the country which we loved, and we only wish to wander on the prairie until we die.

The vast empire that had once been scarred only by the hooves of Indian ponies was indeed shrinking. Settlers would build homes, then communities, then cities, as they staked their fortunes and their future upon the harsh terrain of the brush country. Some prospered. Some were left in ruin. Some were too far away to be protected. Others were just too stubborn to quit and too tough to die. The land was as precious to the pioneers as life itself.

So many homesteaders — even the settlements themselves — shouldered the torment of the plains alone. Droughts choked them. Blizzards trapped them. Sandstorms blew away both soil and seed. Those roving bands of Indians never let any of them out of their sight. Rangers patrolled the plains. Soldiers spent their days in the saddle, searching for marauders who would strike — then vanish in the open prairie that hid them. Too often, the soldiers came too late to save a settlement under siege. Mostly, they didn't come at all. Texas was simply too big, its borders too wide, for those lonely homesteaders to find or even expect military protection against the savages who watched from afar with the paint of death splashed brilliantly upon their faces.

Men took care of themselves. Sometimes it wasn't enough.

In Houston County, neighbors built strong log cabins close together, creating a fortress — a wooden wall of safety between themselves and the Indians. It was as useless as cornstalks. The raiding party slipped upon the cabins by the light of the moon, carrying only tomahawks and butcher knives. They left behind seven slain women and children, all lying in the darkness. The blood had splattered so thick that it extinguished the flames in the fireplace.

Another band attacked a man named Harvey as he worked in the fields. He ran for the gun that had been hung above the doorway, but a bullet stopped him short. His wife crawled under the bed, but a brave discovered her and dragged her fighting and scratching and screaming out in the yard. He killed her then ceremoniously cut out the woman's quivering heart and laid it upon her breast. Neighbors found the Harveys' ten-year-old son with twenty bullet holes torn into his small, frayed jacket. And their nine-year-old daughter was missing.

Along the banks of the Guadalupe River, Indians surprised nine surveyors as they scooped out honey from amidst a clump of bee trees. They died even before they knew that death had found them. When hunters stumbled across the remains of the surveyors, all they found were gaunt, bleached skeletons grinning up at them from out of the bee trees. Only one could be identified. His name was Beatty. He had lived just long enough to carve his name in the root of a tree, just long enough to fashion his own tombstone.

William Adkisson, Daniel Hornsby, and Reuben Hornsby sat fishing in the Colorado River while a small band of Indians crawled up behind them and rammed spears through their backs. One died instantly. Another tried to swim to safety, but a brave paddled along beside him, stabbing an arrow into him until he could swim no more. Only one found freedom.

In 1839, William B. DeWees sat down in Columbus and wrote a friend, stating simply, yet poignantly, the problems faced by those who journeyed west to find a home in a land that didn't want them. He said:

> ... we seem to be surrounded by a foe whose hostility can never be exhausted; I refer to the Indians. Their numbers seem to increase rapidly, and within the past year they have been more hostile, come farther into the settlements, and committed more daring depredations than they have ever done before. They have ever annoyed us exceedingly by coming on to our frontiers, stealing horses and murdering the inhabitants ... The arrival of the northern Indians in their midst seems to have added fury to their breasts already inflamed with wrath and hatred. Now no danger, no fear deters them from venting their rage upon whoever they

find in a situation but poorly protected . . . (the Indians) killed two men, a white man and a negro, whom they found erecting a cabin . . . On arriving at the spot, we found the bodies lying there butchered and mangled in horrible manner; one of them had been scalped, and the whole presented a most ghastly spectacle.

The prairies of Texas lay silent, calm, yet smouldering. They had been baptized in the blood of an unholy war — that seemed holy only to those who went to battle. Men killed for what they wanted and died for what they refused to give away. On those desolate, isolated prairies, white and red warriors alike found a home but no refuge. And they fought to hold onto a fragile patch of ground that would never really belong to either of them.

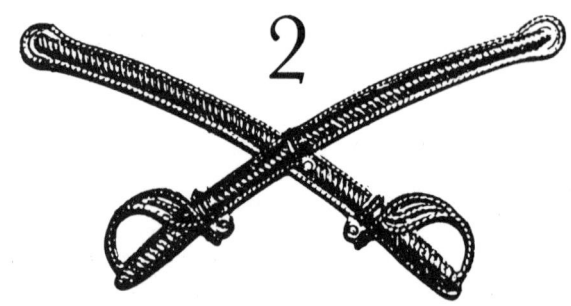

The Guardian of the Prairies

The migration of white footprints crept ever westward. The land beckoned, and stouthearted men could not resist the temptation of owning their own little corner of Texas, even though much of it was bleak, barren, and as friendly as the rattlesnakes that hid beneath its rocks. They trudged on toward the setting of the sun, on horseback and in wagons and on foot. Many grew old in a hurry. Many would never return, and the winds swept down to erase their tracks as though they had never ventured out into the prairies at all.

The Comanches watched with eyes cold and narrow. They rode upon a sanctuary that the U.S. Army had granted them, a place of refuge, and they rode without fear, and they left an ashen trail of death and sorrow behind them. Settlers had been warned against intruding on Indian lands, but they refused to listen to the admonitions of the military authority. They merely saw what they wanted, and they grabbed it, and, often, it passed through their hands like sand in an hourglass.

The problem worsened. And the stain of blood spread far and wide upon the plains.

On March 6, 1847, Captain James H. Ralston, the assistant quartermaster of volunteers at San Antonio, wrote Colonel G. Croghan, expressing his concern over the reasons behind the brutal prairie warfare:

... The Indian Tribes residing within our boundary have heretofore been regarded as dependent nations, with whom we may treat, entitled to our protection; they are entitled to protection alike against the encroachment of individuals, or the usurpation of States.

Since I have been on these frontiers, I have been forcibly impressed with the conviction, that most, if not all, the aggressions, complained of against the Indians may be traced immediately to the improper encroachments by white men on Indian rights; the stupidity of white men leads them into the Indian Country to survey wild lands, the Indians, to prevent such encroachments on their hunting ground, murder the surveyors and others found in their territory; thus a continual excitement is kept up on the frontiers, and peace and safety cannot be had till a treaty boundary has been established, beyond which the surveyor shall not be permitted to plant his jacob staff. Such a boundary being established, a very few small frontier Posts, would give security and peace to the whole frontier of Texas ... the United States would be inexcusable for permitting the Indian Tribes to be exterminated or driven out of the state, for the benefit of a few Surveyors and land speculators.

U.S. troops kept a watchful eye upon the prairies, but there was no thought at all of either exterminating the tribes or driving them beyond the boundaries of Texas. Their hands were tied by the bureaucratic red tape in Washington. After all, the Comanches had their sanctuary, their government reservation, and the army must leave them alone. The horse soldiers had not been granted permission to even chase the marauding Indians, much less fight them.

Men died, both white and red.

Early in 1840, it was hoped that a lasting treaty had been signed with the Comanches who promised to return all of the white women that they had stolen off the plains. In San Antonio, only one captive, Matilda Lockhart, was brought forth. *She's all we have,* the chieftain said. *There are others,* the woman screamed, *hidden back in the mountains.* The voices of the treaty commissioners were raised with bitter accusations. The voices of the warriors were raised with indignant denials. A fight abruptly broke out that would stain the council house with blood. The sudden rage left fourteen Rangers and commissioners dead, and thirty warriors had fought their last battle.

The Comanches seethed in their villages, vowing retaliation for the slaughter of their people. Revenge came quickly. Linnville was pillaged and burned. At Plum Creek, a force of two hundred white men, led by a band of Texas Rangers under Captain Ed Burleson, drove the Comanches into a westward retreat, suffering heavily from their losses. The Rangers pushed on, and the miles took them further away from home. Weeks passed, then months. At last, scouts found the Indians again, resting in peace and solitude upon the banks of Las Moras Creek. A thousand head of horses and mules had been herded into the valley. The Comanches had outdistanced their fear, for white men, not even soldiers, dared to venture so far west. And they slept in ignorance as Captain Tom Howard's expeditionary force crawled across the flats and attacked.

An article in the May 3, 1906, issue of *The Brackett Mail*, compiled from eyewitness accounts, reported:

> To say the Indians were surprised puts it mildly; they were panic stricken and fled in great crowds to the west and down Las Moras Creek, many of these being women and children. A strong body of warriors made a stand along the foot of the ridge on which Fort Clark is built but soon gave way before the fatal fire of the rifles. A great deal of noise and confusion prevailed. Settlers and Indians covered a mile or more of country — Indian warriors yelling, squaws and papooses also, and to add to the terror screaming, white men yelled of the scene; the great herd of mules and horses up in the valley stampeded and their running made a great noise like the near approach of a cyclone. In less than an hour, not an Indian was in sight except the dead warriors . . . Among the white men, many were hurt but not one lost his life.

The sore that festered upon the prairie had not been healed with the battle. The volunteer army rode away home, and the Comanches remained to lick their wounds and fight again. Angry, desperate settlers begged the U.S. government for protection. Their pleas, for so long, would fall on ears that were too far away to care about them or their plight. And the Rangers were virtually no help at all. In the decade following the Mexican War, most of their seasoned Indian fighters all rode away, searching for their families, their homesteads, a new life. They were battle-scarred and weary. Only a few remained behind to uphold the standards of the Texas Rangers, but they were poorly paid and sometimes poorly led and unable to defend such a vast stretch of prairie.

The Indians grew brave, then reckless. There was no one to stop them. Even the Wichitas, Tonkawas, and Lipans lashed out at the homesteads which faced the sprawling frontier with few guns to defend them. Men had to fight. It was too far to run.

During the hostile year of 1849, at least 149 of them, their wives and children, fell before the onslaught of Indian raiders. By September, the *Texas State Gazette* quoted one of the settlers as saying,

The Guardian of the Prairies

> I see that the Comanches are still continuing their forays upon the Texas border, murdering and carrying off defenseless frontier settlers who had been granted protection ... They must be pursued, hunted, run down, and killed — killed until they find we are in earnest ... they must be beaten up in all their covers and harassed until they are brought to the knowledge of ... the strength and resources of the United States.

The next two years brought neither peace nor relief. The Wichitas, Tonkawas, and Lipans rode with grim faces across lands where they had traditionally hunted and felt at home. The terrain had grown sterile, as bleak as the skull of a buffalo bleaching in the sun, even ignored by buzzards who had no more bones to pick clean. The tribes knew their game had scattered, or had been slaughtered and left to rot, and winters suddenly pushed them onto the threshold of starvation. They were neither strong enough nor foolish enough to venture out onto the plains ruled by the powerful Comanche warlords. Yet they and their people were hungry. So they plundered the farms and homesteads and lonely dugouts that wedged themselves between caliche and tumbleweeds. It was their only chance for survival. And they took beeves and horses and, sometimes, a life or two.

By 1852, the U.S. Army had decided, at last, to establish a chain of forts across the western fringe of Texas and bring a measure of protection to a harassed land. One post would be nailed down to a rocky ridge of limestone that curved along the shoreline of Las Moras Creek, down in the isolated brush country that swept on toward Mexico and the Rio Grande.

For centuries that bubbling water of Las Moras had been an oasis in a dry and thirsty land. The Lipans and Comanches and Kickapoos had all fought — and many had died — for the rights to the spring that beckoned great herds of buffalo to the valley of rich grass. Comanche teepees had rotted away. The brown huts of the Lipans had been threshed and winnowed away by the seasons and the winds. For the nomads followed where the buffalo led them, and always they came back to the cold waters of Las Moras.

A Captain Flores had found the artesian springs in 1732, as he waged a campaign against Indians who had a bad habit of attacking and ransacking the ranches that sprawled among the foothills around the outskirts of San Antonio. About three decades later, Captain Philipe de Rabago looked with satisfaction toward Las Moras, when he began trying to develop an easy trade route between Mexico and the stone mission of San Saba. It would, he reasoned, be the perfect site to camp overnight, a refuge for the weary travelers who had spent too many days choking on too much dust as they labored between the boundaries of two nations. In January of 1762, the captain, along with Father Ximinez and thirty soldiers, delayed their journey for a few days as they rested beneath the great live oaks that sheltered the passageway toward the muddy Rio Grande. Rabago was impressed. He even seriously considered establishing a permanent mission to bring the foreign words of Christianity to the three tribes of heathens who worshipped only the land and the sun that warmed them and the buffalo that kept them fed. But, alas, the good Father Ximinez didn't care for the hard rock soil around Las Moras at all. He wanted a mission all right, but he wanted to see it rise up in stone and adobe above the headwaters of the Nueces River. Father Ximinez got his wish. Yet his church beside the molasses flow of the Nueces died early and faded quickly.

The feet of the *padres* came often to Las Moras, as they trekked out of Mexico and moved on across the brush country to the small, frail missionary altars that clung to the riverbanks of the San Saba.

Others made it a point to stop at the springs as their wagon trains lumbered west out of San Antonio and rocked unsteadily along through the dust and the grit of the old El Paso Road. But no one ever came to Las Moras for long. They feared the Indians who waited as ghostly sentries back beyond the wind-chiseled ridges of the hill around them. They knew that the tribes had no use at all for the *padres*, and the warriors respected and trusted the traders even less. So on the travelers came, and they drank from the spring waters, and they left before the sun had had time to dry their faces. They took chances when the odds were so greatly stacked against them upon this wilderness crossroads that drove far into the heart of the Indian country. It was always an antagonistic, treacherous stretch of dirt highway, packed hard by the pounding of feet, rutted deep by wagon wheels that sometimes started west but never arrived. It beckoned only to those strong enough and defiant enough to turn their backs on civilization and leave it far behind. Yet, as they moved westward, they brought civilization with them, scattering it — as the wind blows the thistle — upon the plains to take root in a sun-scorched soil that was tolerant only to the cacti and the scorpion.

During their war with Mexico, U.S. soldiers had often camped beside the artesian waters of Las Moras, spending their last night on American ground before unstrapping their rifles and marching southward into a foreign land of tempest and tamales. It has even been said that Major John B. Clark paused beneath those live oaks before leading the U. S. First Infantry into battle. Some later re-

The Guardian of the Prairies

ported that the major became quite enthusiastic when he talked about the possibility of establishing a military post around the springs that he never really expected to see again.

By 1852, the Indian uprisings had become commonplace, yet they could not dam the flow of wagons west. To men, the only thing cheaper than the land that awaited them was life itself, even their own. Dirt was a lot more valuable to them than blood. Most had never owned any parcel of earth before. The government realized that it would not be able to stop the settlers who were reaching for unplowed land, and the wealth and independence that it promised them. Soldiers could not effectively prevent anyone from venturing into Indian territory, regardless of the treaties, regardless of the solemn oaths made so many years before. So the army might as well go ahead and protect those of its own flesh and blood. It would not stand against men who trespassed, not when the soldiers had a choice, and they did. Instead, they would level their rifles at the savages and push them on until the land gave way to the sea.

The springs of Las Moras Creek were chosen as the location for one link in the government's chain of frontier forts simply because an advance scouting party liked its accessibility, its strategic location overlooking the San Antonio–El Paso Road, and, most of all, its constant water supply. Major Smith, of the Sixth Infantry, was also aware that an old mission and the small, ragged ruins of the Dolores settlement huddled together just a few miles down the road. For years, those patched, weathered buildings had become an important rendezvous point for buffalo hunters, freighters, traders and frontiersmen who had successfully forged their way beyond the western bounds of civilization. A military post on Las Moras Creek would be a sanctuary for them, and, because of its nearness to the Mexican border, the fort would form the southern cornerstone of the Federal defense line that stood guard over the wilds of the entire frontier.

Major Smith, who had ridden into the brush country in search of the site for a permanent fortress, noted that the landscape had a healthy elevation, and he believed that the scarcity of vegetation would undoubtedly cut down on those wanton malaria attacks. He made his recommendations, and the army immediately began making preparations to dispatch troops to a little isolated post that would be known, at least for a time, as Fort Riley, named in honor of the commanding officer of the First Infantry.

In June 1852, Major Joseph Hatch Lamotte spread his Companies C and E of the First Infantry along Las Moras, backed by an advance and rear guard of the United States Mounted Rifles. Lieutenant Sells of California, glanced again at his orders. They were simple and they were direct: "Build a Fort — Build it to last." He would not fail. The terrain had been obviously blessed with good timber and rock, and he sat down back amongst the trees and slowly began developing a blueprint for the guardian of the prairies.

A month later, on July 16, 1852, that little patch of untamed Texas was officially labeled as Fort Clark, a tribute to Major John B. Clark. After all, it had been the major who spread the word on the viability of Las Moras Creek as the logical choice for a military site. He had marched off toward Mexico with the First Infantry, striding boldly into the middle of a war that would take his life on August 23, 1847. Major Clark had been right. He never would see the artesian springs beneath the live oaks and mulberries again. An honored military career that began in 1813 was at last at an end. But his name would not be forgotten, not as long as the flag flew and the cannons roared across the prairies from Fort Clark.

Two weeks later, on July 30, Lieutenant Colonel D. C. Tompkins sat down with Samuel A. Maverick and signed a formal military lease for the land upon which the fortress would be built. According to the agreement, the military would control the property for only twenty years, with the government paying the liberal Mr. Maverick fifty dollars a month. It was a premium price in a country where land was usually worth only five cents or ten cents an acre. Maverick — a lawyer, legislator, and signer of the Texas Declaration of Independence — was a shrewd businessman. The headright certificate for all his acreage had been placed at the head spring on Las Moras, and it was important to him. After all, he had discovered the choice location on an expedition that had almost killed him. Food had dwindled, then run out. Maverick survived by eating roots and grass seeds, berries and the meat from skunks that he had managed to trap. During his negotiations with Lieutenant Colonel Tompkins, the old legislator agreed to let soldiers cut the timber that grew alongside the creek, and meandered back into the dust-blown foothills. They would, he reasoned, need all the lumber that they could find.

When those first two companies of the First Infantry came trudging down to the artesian springs, only one building rose up amidst the oaks, and it was a derelict, fashioned by the hand of a settler who abandoned his cabin and his hopes at about the same time. It had been constructed of cedar poles set perpendicular into the ground, and a thatch roof tried to shut out the wind and the rain and failed at both. Tents surrounded it, and those troops who had been led into the brush country by Captain W. E. Price were turned over for the next two years to Captain J. H. King. Under the direction of

Lieutenant Sells, work began on the construction of a bakery, the adjutant's office, and a guardhouse. They would only be small and temporary, but those buildings would form the cornerstone for Fort Clark. The soldiers settled down to fight for survival upon a harsh and unforgiving land. The Indians became the least of their worries.

When Brevet Second Lieutenant Zenas Bliss rode into the country fresh out of West Point, he noted that "there was nothing but sand, cactus, dense chaparral and poor grass to the end of our journey." Bliss, ever alert in an alien environment, also reported that he observed mustangs, quail, and "Jack-ass Rabbits" running rampant through the prairie that rolled on endlessly before him.

In time, service at Fort Clark meant an assignment into no man's land, a place to forget and be forgotten. It was hard duty upon a hardscrabble piece of land. Men grew tough or they grew old. Each trooper had been issued a horse and told to care for it as though his life depended on the animal. Sometimes, it did. One cavalryman recalled, "For days at a time we did not unsaddle." The surrounding countryside molded and shaped soldiers in its own image, sometimes bitter, always blistered, strong enough to fight and mean enough to look for the chance.

A reporter for an Eastern newspaper, *The Spirit of The Times*, strayed out into the Texas frontier in the years right after Fort Clark had staked its claim in the soil of Las Moras. As if the facts weren't bad enough, with great inaccuracy and misinformation, the correspondent wrote:

> The cattle are not the sole occupants of the prairie by any means. Droves of wild horses are not infrequent, and deer are in countless numbers. The small brown wolf or cayeute [*coyote*] is quite common, and you occasionally get a glimpse of his large black brother. But Texas is a paradise of reptiles and creeping things. Rattle and moccasin snakes are too numerous to shake a stick at; the bite of the former is easily cured by drinking raw whiskey till it produces complete intoxication, but [for] the latter there is no cure. The tarantula is a pleasant institution to get into a quarrel with. He is a spider with a body about the size of a hen's egg and legs five or six inches long, covered with long coarse black hair. He lies in the cattle tracks, and if you see him, move out of his path, as his bite is absolute certain death, and he never gets out of the way for anyone. Then there is the centipede, furnished with an unlimited number of legs, each leg armed with a claw and each claw inflicting a separate wound . . . The stinging lizard is a lesser evil, its wound being likened to the application of a red hot iron to the person; but one is too thankful to escape with life to consider these lesser evils any annoyance. But the insects! Flying, creeping,

jumping, buzzing, stinging, they are everywhere. Ask for a cup of water, and the rejoinder in our camp is, 'Will you have it with a bug or without?'

The prairies around Fort Clark became a curiosity. They seemed barren and lifeless, yet so much lived upon them. The distance between landmarks was overwhelming. The miles staggered men, then the heat broke them. Soldiers were afflicted with a great loneliness that could not be quenched. They were stuck along a limestone ridge — latitude 29 degrees 17' north, longitude 23 degrees 18' west — one hundred twenty-five miles west of San Antonio and forty-five miles north of Fort Duncan, tucked away on the Rio Grande at Eagle Pass. Somewhere beyond the far mountain range, the eyes of the Comanches, the Lipans, the Kickapoos were upon them, watching, waiting to see if the army could stand up to the land or buckle beneath the agony of its solitude. The land worked on men like a thorn beneath the skin. It was irritable at first, nothing more than a nuisance. Then it penetrated deep, and festered, driving some men mad and others to drink. Soldiers got to know the land well, even if they never understood it. The prairies were both fascination and frustration.

In November 1852, a writer believed to be an army officer in the lower Rio Grande, wrote in *The Spirit of the Times:*

> Throughout this vast region the beautiful prairies are interspersed with patches of chaparral of unequal extent and irregular shapes which often give to the scene a curious and picturesque appearance. This chaparral consists of thickets of naked thorns of several species, so thick, tangled, and impenetrable, as to laugh the best cultivated hedgerows to scorn. The bushes are leafless, the arms and thorns as white as if painted, and they glitter in the sun. Attached to the stems and scattered beneath to the depth of several inches, are myriads of shining white snail shells which render the ground almost as white as if it were shrouded with snow. Interspersed among these chaparrals are various trees of other kinds, such as the mesquite-ebony, wild briar, cabbage-wood, and numerous others, valuable for fire wood, fencing, and other mechanical and domestic purposes. When these thickets are cleared away, the ground is exceedingly fertile and easily tilled . . .
>
> In these chaparrals are found countless numbers of rabbits of the ordinary gray species as well as a large gray rabbit, much resembling the English hare in shape but far larger with enormous ears; the true zoological name I do not know, but they are vulgarly called the 'jack-ass rabbit.' They are fine eating, and I have shot them weighing fifty *[sic?]* pounds. Vast flocks of wild turkeys,

some of them very large and all of them fat, inhabit these forbidden haunts. Quails, pigeons, paisanos, and it must be added, no small quantity of rattlesnakes and tarantulas find here a safe and inviting abode. In all parts of this region, deer in fine condition abound, also the peccary, or Mexican hog, one of the most game-blooded animals that exists. They will fight anything, man or beast, and some amusing stories are told of their driving hunters up a tree and there besieging them for hours. We have also a peculiar bird denominated the 'chachalacha' about half the size of an ordinary game cock ... In its native state it is wild and shy, but when caught is easily domesticated and becomes especially fond of those who feed and camp it. At daylight in the morning, whether wild or tame, they commence a furious reveille, repeating in a loud, dismal tone, a chant, from the sound of which they derive their name. This is prolonged for about half an hour — the woods all around you appear to be alive with these invisible songsters, when suddenly they stop and not another sound breaks from them during the whole day. The chachalacha will crossbreed with the common game fowl and produce not only a beautiful bird, but one of the greatest value of its game qualities. Their crosses are a little under size, but in spirit, endurance, activity, and vigor, they are unmatched. They are the best fighting cocks on earth. This is no fancy sketch; they have been tried frequently and never were known to skulk or yield; like the Old Guard, they can 'die, but never surrender.'

... We have likewise another peculiar bird, called the 'paisano' which is deemed of great value by the Mexicans and Indians on account of its hostility to the serpent tribe. It ... can run as fast as a fleet dog. Whenever one of them discovers a rattlesnake or any other serpent, no matter how large, it commences a fierce cry, which summons to its aid all the paisanos within hearing. They begin to run and fly about the snake in a circle, crying and chattering all the time till their victim becomes confused, when, quick as the lightning's flash, one of them and immediately others make a dash at the eyes of the snake, and with their sharp unerring beaks, he is blinded in a moment. He then falls an easy prey to the united prowess. These battles are of frequent occurrence and are described by spectators as interesting in the extreme.

For four years, the troops who occupied the grounds of Fort Clark fought a strange kind of war for survival with the land itself. The Indians, other than occasional harassment, weren't that much of a problem. But then, the Indians, being of sound mind and body, weren't that eager to stand face to face against the strength and the firepower of the First Infantry, backed by the quick striking force of the U.S. Mounted Rifles. They didn't mind sneaking into the encampment in the dead of night to run off with a few horses. However, dying in an all-out war was not part of their plans at all. They conceded the thorny chaparral and the scorpions beneath Las Moras Mountains to Fort Clark and moved quietly out into the prairies just beyond the quick reach of the soldiers. They waited for the heat and the boredom and the acrimonious nature of the country itself to whip the army for them. The Comanches, Lipans, and Kickapoos had learned long ago that they could never tame the land that gave life then took it away. It was too austere, too insolent, too obstinate to ever be conquered, a citadel of hard rock and bleached earth and prickly limbs where only the horned toad and rattlesnake would ever really feel at home. The Indians did not want to possess the land. The white man did. Yet the Indians would kill, if necessary, would kill to keep it free and open to them. They would die so men could build fences. The soldiers were handed the duty of keeping the peace in a country where the peacemaker was never blessed, only cursed by savage and civilized alike.

For a time, it seemed that Fort Clark did not really believe that it had a permanent place upon that harsh western landscape. There was no fortress at all, only a cantonment, a village of tents buried amidst the grit and grime of a foreboding frontier. However, by August 1, 1853, W. G. Freeman, as part of his inspection tour of the Eighth Military Department, reported that Fort Clark was at last "engaged in constructing quarters (some of which are nearly completed)." The soldiers had joined the army to fight. They became common laborers, slowly piecing together an acceptable fort out of stone and mud, though both were bad about eroding away. Only the mail service, riding hard out of San Antonio, kept them in touch with the life they had left behind.

The heat seared their souls. The sky was unmindful of their toil and trouble. During 1853, only 29.4 inches of rain bothered to descend upon the live oaks and mulberries that clustered around the artesian waters of Las Moras. A year later, only 14.42 inches of rain fell. The drought dried the grasses and shrank the creeks. Men suffered but could not leave. Yet others followed the weary, miserable footprints that the army had worn into the sun-baked soil. By 1852, Oscar B. Brackett had established a supply village for Fort Clark, and he called it, with a certain amount of pride, Brackett. His security was the First Infantry. His lifeline was the stagecoach that came rolling in from San Antonio and heading out for El Paso, loaded with the brave, or maybe the foolish, who would find that those long, blistered miles tormented them as much as the Indians.

For the first few years of its existence, Fort Clark battled the elements and, at least, stood its ground. In 1855, however, a new threat rose up to distress the army.

The Indians had grown restless. Their anxiety over the coming of the soldiers had lessened. Arrogance replaced their fear.

They attacked.

It was a bold move that greatly surprised the troops and cost them dearly in horseflesh. The U.S. Mounted Rifles found themselves afoot. And the commanding officer swore that he would not suffer such a loss again.

He ordered the construction of a corral and breastwork just below the springwaters. When night began casting its long shadows across the creek, all the horses and cattle were ceremoniously herded together and driven into the stone enclosure where armed guards kept watch from behind a protecting wall, their guns resting in portholes that had been cut into the masonry.

The soldiers waited.

The Lipans chose to test them.

Hidden by the darkness, a small band of the raiders, led by Chief Thurino, crawled through the thorns and positioned themselves in the sand and rock beside the breastwork. The Indians had kept a close eye on the soldiers, not making their move at all, until only six of the guards were left to oversee the livestock. On that night, six were enough.

One brave reached the main gate and thrust it open. Other warriors raced screaming into the corral, trying desperately to stampede the horses out onto the prairies. The silence of the night was broken by a shrill, yet thunderous, cacophony of shouts and gunfire, cries of violence and shrieks of agony.

The alarm had been sounded.

The six guards stood firm. Smoke hung heavy above the corral as other soldiers ran madly down from the camp itself. The raid had begun quietly, and it ended the same way — with four members of the Lipan raiding party lying dead amidst the frantic horses they had come to steal. For the moment, Fort Clark had proved its strength upon the prairie. The Indians moved uneasily on further west as their foothold in the brush country began slipping away. They would send no more to be buried beneath the mulberries of Las Moras Creek.

Slowly the fortress continued taking shape. A guardhouse was hastily built on the earth about 150 yards south of the corral. By 1857, a headquarters building, hospital, and a bakery had all been created out of that soft native stone from a nearby rock quarry. Fort Clark had forcefully made its presence known. Settlers at last felt safe and maybe even bold. They moved out of the protective reach of the army's guns and pushed on into new territory. The soldiers could no longer be content to merely rise up against any Indian threat that might be riding their way. Now they had to venture out and look for trouble. Most of it, to their chagrin, had escaped across the Rio Grande and fled to the sanctity of Mexico. The tribes taunted the troops that patrolled the chaparral prairie of the brush country. They camped on land where the U.S. Army could not chase them, sneaking back across the border to strike lonely, isolated homesteads, then retreating to foreign soil as quickly as they had come. Theirs was a hit and run kind of war, and the men of Fort Clark found themselves hunting prey that were as elusive as ghosts in a vast, empty land that could swallow them up sometimes without a trace.

On one summer day in 1857, a Fifth Cavalry patrol stumbled across a raiding party, fighting three different battles with the same band of Indians before the sun dropped behind the mountains and the darkness blotted their escape. A year later, troops chased Comanche warriors for three hundred miles before finally losing them beyond the narrow, muddy waters of the Rio Grande. The soldiers quietly set up a trap to catch any of the Indians who might return, but no one came. The mountains held them close and would not give them up. Several months later, a patrol from Fort Clark so thoroughly routed an Indian raiding party that the braves rode wildly for ninety miles before they even dared to pause long enough to make camp.

John P. Prichards, in Company C of the Second Cavalry, sat down in January of 1857, and wrote his brother of the man who led him on such skirmishes across the prairie:

> The Capt. name is James Oakes . . . he is a very fine little man, he is one of the greatest Indian hunters in the U.S. Army. When ever he goes out on a scout he never comes in without he brings some scalps and ponies and mules. C company has killed (more) Indians since we have been in Texas than the whole regiment put together. Texas is wild and barren country. There is not much else in it but mexicans, Indians and other wild animals such as bear, panther and what is called the California lion. It is one of the most savage animals I ever saw.

The Indians, as always, remained a nuisance. Although they were no longer a real threat to the post itself. The cavalry of Fort Clark certainly didn't defeat them, but the horse soldiers did keep the warriors on the run. Settlers felt much better about their chances of survival simply because the rifles of the fortress were aimed in their defense.

On the outer fringe of the rolling prairie, the dry arroyos and chalk mountains seethed with danger.

Fort Clark, according to Lydia Lane, the wife of a lieutenant, "was a pleasant post." She and her husband had arrived in Texas in the mid 1850s, then traveled on to Clark from Fort Inge after it had been abandoned. Mrs. Lane wrote,

> The change was very aggreeable to us all, the garrison being a large one, with a number of officers and ladies.
>
> A funny little house had been put up for us before we arrived, all the quarters for officers being occupied. The walls were built of green logs with the bark left on them, and they were set up on end, — not like the usual log-cabin. The Mexicans call a house of that kind a "jacal" (pronounced hackal). The walls were seven or eight feet high, and supported a slanted roof. There was really but one room in the house, with an enormous chimney, built of stone, in the middle of it. The spaces between the logs were chinked with mud, or plaster, perhaps . . . We had no ceiling, — nothing but the shingles over our heads, through the long, hot summer . . .
>
> Our little house was so far from the other quarters, I think the Indians could have crept in upon us, taken our scalps, and ridden away, without being molested. Nothing troubled us, however, but the field-rats and mice, which were in numbers when we first occupied the house. They came into the room round the walls, where the boards of the first floor were scooped out to fit the upright logs of which the house was built. All being green at first, they dried during the intensely hot summer and very soon the floor and walls were far apart, so that the rats and mice came and went without ceremony. We saw a rat drag a small bottle of sweet-oil from one side of the room all the way across, and down under the floor on the other side.
>
> The rats and mice were bad, but we found a tremendous snake on the mantelpiece, and that was much worse. I was just about to retire one night, when we heard a suspicious rustling among some papers, and there he was, moving cautiously among them; how it ever got up there we could not imagine. I fled out of doors, while husband killed it with his sabre. Another large one was killed in the brush at the end of the porch. Sometimes a skunk would pass the house, but never very close. He is a beautiful little animal to see; but distance lends enchantment in his case.

For Lydia Lane, the critters were her torment. For John Prichards, it was the weather. He told his brother:

> We have had no cold here this winter of any account. There is

whats called a norther here once in awhile but it don't last over twenty four hours and after that it gets warm and then you can go in your shirt sleeves for a week or two and then comes a[nother] norther. I don't like this country very well.

Life wasn't particularly easy at Fort Clark during the late 1850s.

But it was bearable.

3

Surviving in a Selfish Land

The land had no friends. It was hard and sometimes mean, and it had no tolerance at all for those who came to stay. It broke them and buried them and swept away their memories from the burning face of the land. Albert J. Myer rode silently across the prairie, virtually lost amongst the "clumps of brambles" and "pigmy tree" that grew alongside the trail. He marveled at the "remarkable cacti — every variety from the Turks Head to the Strawberry," and he thought that the "maguey with its fleshy leaves" somehow symbolized the countryside that sprawled around him. He wrote: "You can gather it here at any time and the Mexicans will roast its roots and try to persuade you in spite of your own teeth and tongue that it is good to eat." So many looked upon the land as their Canaan. So few would be able to survive the hardships that grew with the prickly pear beside their homes.

It was in March 1834, when Dr. John Charles Beale and his Spanish wife, Dolores, turned away from the Texas coast and began looking westward. They had brought with them forty-five families from New York, sailing to the shoreline of Aransas Bay, hoping and praying for a patch of ground that they could — at last — claim for their own.

Dr. Beale had advertised that he would find a "dream place" for the settlers, and they trekked for two months across a strange and unmarked country, free and deserted, empty and awaiting them. Amongst the chaparral, just north of Las Moras Creek, they stopped for a final time, too weary to move on, nailing the stakes of long homesteads into a dry and rocky soil.

Dr. Beale wrote with great enthusiasm: "To our great joy, we camped on the future site of Villa de Dolores."

He was home.

At least, he thought he was.

Work began on a small fort and fields were plowed. Even a small shallow ditch was dug for irrigation. A village took shape among the crude huts. March 25 became a day of celebration. Dr. Beale wrote:

> Immediately after breakfast, everything being prepared we marched in procession to the site of the church. The Commissioner and myself, with the Mexican flag, leading the way; next to us were two master masons, one carrying a stone and the other a portion of mortar. On arriving at the place, we found that a small part of the foundation of the church had been dug; one of the masons prepared the bed, and I then laid the first stone of the Villa de Dolores; a bumper of wine was then tossed off to the prosperity of the new town, amidst cheers and repeated firing of guns. We now proceed to swear allegiance to the Mexican Republic, which was done first by myself, and then by all the rest of the colonists ... The day was beautifully fine, and everything passed off with the greatest order and good humor.

Dr. Beale's "dream place" was barren and desolate, as hostile as the spikes on the cacti that sprawled across open fields. The heat seared the land and scorched the thirst that rose up like a prayer in the throats of men and women who had ventured out onto a frontier where they didn't really belong. Scorpions taunted them. Vinegaroons ran rampant at their feet. But those families did what they could to chase misery away from the wasted prairie. They cleared the land for a gristmill and a sawmill. Homes were pieced together. Seeds were scattered upon harsh fields.

The dream died.

A drought sucked the plains dry, and the sky refused to rain. The irrigation ditches were thick with dry dust. Crops withered away. The Comanches left the prairies scarred with death and ruin. News finally reached Villa de Dolores that a Mexican General named Santa Anna was marching his troops across the Rio Grande, intent on crushing the rebellion that had swollen up within Texas. He would walk over anyone who stood in his way.

Fear gripped the village.

The families had faced hardships without complaint. But none were ready to die for a cause that belonged to someone else. Early one morning, the settlers, in a virtual panic, quickly loaded their wagons and rode away from Villa de Dolores. Dr. Beale was out of town, and there was no one strong enough to hold them or keep them together. The colonists had one thought in mind as they threaded their way through the bramble bush that had never befriended them anyway.

Escape.

Some didn't make it.

The Comanches caught them and left them bleeding on a land that had no one to bury the dead or mourn them when they were gone. Beside a large lake, on the frantic road to Matamoros, Mrs. Harris, Mrs. Horn and her two small sons were taken prisoner by the Indians. For a year, the women were dragged across Texas and Mexico by the nomadic tribe before New Mexico Comancheros finally paid their ransom and set them free. The fate of Mrs. Horn's children was never known. Perhaps they had been adopted into a Comanche family. Perhaps they died in captivity.

Back near the springwaters of Las Moras, the village of Dolores aged and crumbled. No one came anymore. No one cared.

The land was selfish. It tolerated the Indians only because they never stayed in one place for very long. There was no permanence in their footprints. They had no inclination to own the ground around them, only to survive upon it for a time. When the soil and the rivers could no longer feed their hunger, they would be gone to another place, and the land would forget them.

To the settlers, land was a valuable asset. It was something to possess and keep. It belonged to them, or so they thought, and most would fight to hold it. But only the strong would ever inherit the earth. The weak never lasted long enough for their roots to grow very deep. The cowards never came at all.

Dr. John Charles Beale had tried to tame — or at least civilize — a patch of the prairieland and failed. The U.S. Army would follow in his footsteps, slowly passing the ghostly relic of Villa de Dolores, as its soldiers patrolled the plains beyond Fort Clark.

The military had force. It had the strength of numbers and the firepower to protect the western terrain of Texas. The army would learn — as Dr. John Charles Beale had learned — that the land never fought fair, and the land gave no quarter.

Disease stalked the military posts that stood sentinel around the southern and western frontier of Texas. It was, perhaps, the most deadly enemy that many of the soldiers would ever encounter.

During one epidemic at Fort Brown, for example, almost one-fifth of those struck down by yellow fever died, even though physicians tried to ease the suffering with "general and local blood-letting, [and] calomel, combined with quinine . . ."

At Ringgold Barracks, on the left bank of the Rio Grande, assistant surgeon Israel Moses fought cholera with "quinia, aided by wine, brandy, and other stimulants." Still, five of the troopers died.

Cholera, as Surgeon N. S. Jarvis of Fort Brown pointed out, spread with the force and the fury of a forest fire after the first spark had been ignited and fanned by strong winds.

He reported,

> Its ravages have proved equally fatal along the whole valley of the Rio Grande, passing with fearful and fatal strides from one town to another, some of which [are] nearly depopulated, and making sad ravages in the different ranches that extend along the banks of that river. At Camargo, it is supposed to have carried off one-third of the inhabitants, and the town of Roma was nearly abandoned by the inhabitants to escape a similar fate.

At Fort McIntosh, assistant surgeon Glover Perin learned from a Catholic priest that scurvy could effectively be treated by the juice from species of the agave plant. He was skeptical. He preferred using lime juice as a remedy, just as his old, faded medical books had instructed him to do. Alas, he could find little of it, so out of desperation, Perin turned finally to the aloe vera, having no other alternative but to trust the priest and his word.

The assistant surgeon would report:

> Private Turby of Company G, 1st U.S. Infantry, was admitted into hospital March 25, in the following state: countenance pale and dejected; gums swollen and bleeding; left leg, from ankle-joint to groin covered with dark, purple blotches; leg swollen, painful, and of stony hardness; pulse small, feeble; appetite poor; bowels constipated. He was placed upon lime-juice, diluted and sweetened so as to make an agreeable drink, in as large quantities as his stomach would bear; diet generous as could be procured, consisting of fresh meat, milk, eggs, etc.; vegetables could not be procured. April 11th. His condition was but slightly improved; he was then placed upon the expressed juice of the maguey . . . April 17th. General state very much improved; countenance no longer dejected, but bright and cheerful; purple spots almost entirely disappeared; arose from his bed and walked across the hospital unassisted; medicine continued. May 4th. So much improved as to be

Surviving in a Selfish Land

> able to return to his company quarters, where he accordingly went
> . . . May 7th. Almost entirely well.

Glover Perin continued:

> Eleven cases, all milder in form . . . were continued upon the lime-juice . . . They exhibited evidences of improvement, but it was nothing when compared with the cases under the use of the maguey . . . At this time, so convinced was I of the great superiority of the maguey over either of the other remedies employed, that I determined to place all the patients upon the medicine.

No longer would men curse the thick, dagger-edged leaves of the agave plants that rose up between the rocks and alkali dust of a barren land. The knowledge of the aloe vera plant would be a godsend to all of the frontier forts when scurvy began to knock their troops from the ranks of able and stouthearted men.

As Perin wrote of the maguey:

> As it delights in a dry, sandy soil, it can be cultivated where nothing but cacti will grow; for this reason, it will be found invaluable to the army at many of the western posts where vegetables cannot be procured.

The land hurt.

Sometimes it healed.

Men grew restless upon the prairie, tormented by sickness and loneliness and the threat of death that lay beyond every canyon rim. Home seemed so far away as soldiers dressed themselves in tunics and hot canvas trousers, then shouldered nine- and one-half pound weapons and trudged in ankle-deep dust out beneath a sun that blistered the ground around them. For some, not even the magic of the maguey could help ease the pain that burned within them.

They ran.

The prairie swallowed them up, and the winds swept away their footprints as though they had never ventured out upon the land at all.

Five officers and 109 enlisted men had journeyed from Ringgold Barracks to the banks of Las Moras Creek to establish Fort Clark. For some, it was as though they had been exiled to an alien country. Duty had become a form of punishment, and Mexico lay so near to them. Freedom sprawled just beyond the Rio Grande. By July of 1852, a first lieutenant had gone AWOL. In October, the captain sat down, gave himself an indefinite leave, and was never seen again. His successor only lasted for two months before granting *himself* an indefinite leave as well, and walking away from his command. Neither captain ever received a reprimand for negligence or desertion. Instead, the army simply promoted both and transferred them to Washington, D.C. They never got the word. They never arrived.

Only an inexperienced Second Lieutenant Hudson was left in charge of a dying post that had stumbled and fallen on the threshold of extinction. He looked around and found only seventy-five men left in the garrison. He was their last officer, thrust into an unfamiliar role. He promptly got in touch with the commander of the Ringgold Barracks, then sat back and waited for him to send help.

A colonel rode up from the Rio Grande Valley, swaggered across the wilderness parade ground of Fort Clark and immediately began to straighten out the problems that plagued the ramshackle outpost. He knew men, and he was a stern judge of character. As he wrote of one man, ". . . [he] left post with a certificate of disability, but I think he lies."

Leaving wouldn't be so easy anymore. The colonel cracked down, and almost overnight the strength of the enlisted men rose to eighty-six. For a moment, the death rattle at Fort Clark had been silenced.

The soldiers at the post still viewed themselves as condemned men who had been banished to a condemned land. Many of them lived in lean-tos or beneath overturned wagons. The fortunate troops had tents that leaked when it rained and were stifling — almost suffocating — when the summer sun hammered down relentlessly upon the prairies, and it almost always did. One military report said of those soldiers who patrolled the western frontier:

> They have generally made for themselves bunks of grass and the branches and leaves of trees, which, raised a little from the ground, protected them from wet and damp . . . The men have subsisted on fresh beef, pork, and bacon — the former being issued once in five days — hard bread, beans and rice, sugar and coffee; they have eaten no vegetables but prickly pear and poke-weed. In September, molasses, pickles, and dried apples were received, and issued in lieu of some portion of the ordinary ration.

The troops were given prickly pear and pokeweed to cure them of scorbutic taint, an affliction which caused livid spots and spongy gums. Later, they suffered from an intermittent fever. The reason? They had eaten, so the physicians said, too *much* prickly pear and pokeweed.

The surgeons who stood watch over the sick and the wounded kept men alive with homespun remedies and as much luck as skill. As one would later admit,

Surviving in a Selfish Land

>   Sometimes they lived, and sometimes they died. Mostly, they lived, because people are so tough.

They had to be tough. Medical supplies were picked up in New Orleans, and it took more than a month for them to reach Fort Clark, perched at the far end of a long ride. Those who suffered a misfortune simply suffered.

One surgeon who did his best to patch up the western frontier of Texas wrote:

>   Fresh wounds require daily dressing to protect them from maggots, where there is no other necessity.
>
>   Sergeant W., bit in the right hand by a rattlesnake, had his wound dressed constantly with lint saturated with liquid ammonia; notwithstanding which, the maggots got into it; they were, however, speedily removed by a dressing of calomel — a never-failing remedy . . . In the case of a German laborer . . . who had been wounded in the knee by a rifle-ball . . . I took off the limb at the upper third of the thigh about 36 hours after the accident, under the influence of chloroform, and never had an amputation to do better . . . He had lost much blood, was exceedingly irritable, and, but for the aid of chloroform or a narcotic, I should have had trouble in controlling him.

But the men persevered.

And they survived.

Only seldom did they pass a day without some kind of skirmish with Indian raiding parties that rode the prairies and scowled at anyone who dared to intrude upon their hunting grounds. For weeks at a time, soldiers were even afraid to unsaddle their horses. They never knew when an attack would come, or if it would come at all. The Comanches left the homes of settlers in flames and in ruin, leaving no one behind to pray for the dead after they had gone. They struck and struck quickly, preying like vultures upon supply wagons that ventured beyond the realm of military protection. The Indians gave Fort Clark a wide berth, watching the soldiers with anger, but with caution.

The soldiers learned to be just as wary.

On one Sunday afternoon, two troopers rode arrogantly out across the hills and to the summit of a small mesa. In the distance, they saw a long, sinister line of dust clouding up from the prairie floor. An Indian war party was thundering across the landscape, and Fort Clark lay in its path. Fort Clark was unaware that trouble was riding its way at all.

The two soldiers turned their mounts and headed hell-bent for leather toward the isolated little outpost, jumping Las Moras Creek, and yelling, "Indian attack! Indian attack!"

That cloud of dust rolled on toward Fort Clark like a West Texas sandstorm, whipped to a frenzy by an angry spring wind. The troops knew immediately that they were outnumbered and saw no reason to make a foolish stand in a battle they could not win. After all, their fort was nothing more than a few scattered tents anyway. They cut their horses loose and hid back beneath the shelter of the brush country, watching as the band of Indians swept past them. That night, the soldiers fanned out and trekked southward toward the Rio Grande. By morning, they had found their stampeded horses and returned to the camp along Las Moras Creek.

The prairies had taught the infantrymen one basic lesson. They had learned patience. They knew that some day revenge would finally belong to them. So they sat, and they sweated. They waited, and hatred throbbed deep within them. Their day would come, and they could erase such humiliation from their minds. Time was on their side. They had plenty of it, and time became the one burden that could turn civilized men into savages.

On a cold January morning in 1855, Companies A and G of the Regiment of Mounted Riflemen stumbled across a band of Comanches. Dr. Albert J. Myer wrote:

> ... an Express came in with the news that the troops who had gone over the road we were to follow had killed a party of seven Indians and captured a little squaw. They came upon them one morning unexpected and the [I]ndians fled into the tall cane and grass which bordered the Pecos River. Fled, however before the whites had fired at random into a crowd of Mexicans with whom the Comanches were trading. One [M]exican was killed on the spot and others were badly wounded. We met them on their way down as we left Fort Clark with an escort and I could not but pity one poor fellow who had ridden on horseback nearly 200 miles with his leg shattered fearfully by a musket ball. As the command came up, the Texas Rangers and the Rifles, spread themselves along the river and the grass was fired: as the flames drove each from his hiding place fifty balls were placed in his body and so seven red-skins were massacred.
>
> Had the Comanches in proportionate numbers surrounded a few whites, the affair would have been styled a diabolical murder; as it was, we must try to look upon it as an execution. The war on this frontier is one of extermination. In the worst sense of the word these tribes are savages. They are devils and the coldest blood must boil at the narration of the manner in which they have treated prisoners who have fallen into their hands, not men, alone,

taken with arms in their hands, for they can but die, but innocent women and children. Orders are now issued to the troops to take no prisoners; to spare no one; to listen to no terms for peace until the race is cowed by their punishment. You can form an idea of the state of the country through which we were to pass.

Fort Clark, by 1855, had become a formidable post, no longer the place that had so thoroughly disgusted a surveyor two years earlier, causing him to write that the morale of the troops was low, the hospital area was merely a tent city, and the hospital conditions were ghastly. As far as the surveyor was concerned, the terrain itself was "harsh," and the weather "abominable." After all, it had taken him three days just to travel the forty miles from Eagle Pass, and he rode away as quickly as he could, leaving behind a map so hastily drawn that it had neither north nor south inscribed upon its stained, wrinkled face.

Dr. Myer didn't find the countryside around Fort Clark so disagreeable at all. He wrote:

> ... we camped at a beautiful stream. I had seen no running water before, in Texas, and to hear it rushing over the little rocks and tumbling into deep basins carried me back to my home, in thought. Here too we saw beautiful fish, so they seemed at least, for I had seen none in fresh water, since I watched them in 'Eighteen Mile Creek' (in New York).

For him, it was a good country, troubled only by the piercing eyes of the Lipans and Comanches whose wrath had deeply shattered the peace and tranquillity of the Trans-Pecos. When Assistant Surgeon Basil Norris visited Fort Clark in September of 1856, he agreed with Myer, reporting that the outpost "occupies a healthy site upon the Rio Las Moras; it overlooks the surrounding country and commands a view of the Rio Grande bluffs, thirty miles to the southwest; and to the east and north, distant mounds and mountain ranges are distinctly visible."

The soldiers, he pointed out, had generally escaped the suffering and the sickness that had plagued so many of the troops stationed within the cordon of western forts, explaining,

> The shallow soil of the prairie supports a scanty growth of vegetation, and the cause of malaria is nowhere to be found except upon the river course ... The eruptive fevers have been fortunately almost unknown; a case of variola, brought by a citizen from El Paso, was treated and cured under our care; and, although left by those who brought him to this point, in the midst of the garrison, the precautionary measures adopted proved successful in preventing its spread among the troops.

Scouting parties faced the daily threat of severe sickness. Their task was an arduous one, their rides long and wearing. They picked up the seeds of disease from those suffocating days and nights when they camped beside the stagnant pools of the bottom lands. The water was bad, but it was all they could find. Thirsty men didn't care how foul, how filthy the bug-infested mud holes might be.

But mostly, at Fort Clark, Norris attributed health problems to errors in diet, sudden changes of temperature, and the abuse of intoxicating liquors.

He reported:

> To the ration with fresh beef, vegetables from the garden are supplied in sufficient quantity to flavor and improve the soup, and occasionally, during the year, purchases of onions, sweet potatoes, etc., are ordered for the whole command.

Better times had settled down around the outpost. No longer were soldiers being forced to fill their bellies with prickly pear and pokeweed to chase away hunger pains.

Norris continued:

> The clothing used by the troops is, in my opinion, too thick and warm for summer.
>
> The quarters, which were heretofore only comfortable during the summer months, are made of pickets driven into the ground, and covered with rudely thatched roofs. They have been recently repaired by weather-boarding from the 'portable cottages' and, with additional chimneys, will be suitable for winter.
>
> The water used in the garrison is drawn from the river and kept in barrels, which are cleaned and replenished every day; it is clear, slightly impregnated with lime, and is, at all seasons, cool, refreshing, and wholesome.

The Indians let the soldiers possess the running water of Las Moras Creek undisturbed. They could find no advantage at all in launching an attack on Fort Clark. To them, it just wasn't worth the cost. All the post had of value anyway were some horses, and the Indians refused to die over a few aging mounts that they really didn't need.

They might *steal* when they could.

From time to time, they might even harass the outpost that lay on the far side of the live oak, pecan, and elm.

But they preferred to roam a wild, open country to the west of the fort, miles beyond the reach of the post guns. On the prairie — too vast to protect with a few horse soldiers — supplies were much easier to take, and so were lives. So the soldiers trailed after them, chasing rumors and cold, forgotten tracks across the Trans-Pecos and virtually all the way to El Paso. What help they received in their quest to keep the Texas prairie safe and free from bloodshed came from the Texas Rangers.

Albert J. Myer wondered if the Rangers weren't, in reality, worse renegades than the Indians. He referred to them as animals, "...rowdies, rowdies in dress, manner and feeling." He wrote:

> Take one of the lowest Camel drivers, dress him in ragged clothes — those he ordinarily wears, as you see him, are altogether too clean — put a rifle in his hand, a revolver & big bowie knife in his belt — utterly eradicate any little traces of civilization or refinement that may have by chance been acquired, then turn him loose, a lazy ruffianly scoundrel in a country where little is known of, less cared for, the laws of God or man and you have the material for a Texan Mounted Ranger ... of whose class some hundreds are at present mustered into the service to fight Indians.

As Myer and his party rode away from Fort Clark, two of the Rangers joined them as escorts, talking about the dozen Indians they had encountered about midnight.

"We hollered and they hollered," said one of the Rangers.

Albert Myer wrote, "As far as we could comprehend both sides were badly scared; both sides ran. It was nothing but two Rangers scaring twelve."

But guns in the hands of friends — even some of the rougher types of Ranger — were welcome as Myer and his expedition rode out toward Fort Davis, pushing on into a prairie empire ruled by the anger of the Comanches. It was a place, he wrote,

> ...where the white man goes now ready always to fight, for he knows that spying eyes are ever upon him and savage war-parties ready to strike at the first negligence ... Danger comes openly on the prairies. There are no trees to hide a foe, and the savages of this country fight on horseback. They are the best horsemen in the world. In battle they gallop around you, throwing themselves out of the saddle and hanging over the horse's side keeping the body of the horse between their enemy and themselves, while only one foot is seen clinging to the saddle. It requires skill to do this alone, but in addition, these warriors manage to keep arrows flying from under their horses necks with a degree of certainty and rapidity that is disagreeable to contemplate.

The Comanches spread more than 9,000 strong across the Southwest, but only rarely did warring raiders number as many as

200. In Texas, they had found wood, good water, and abundant game — but the soldiers stood in defense of those who wanted to take it all away.

According to their chieftains, the Great Spirit had brought them to this land many generations ago, freeing them for all time from the persecution of their enemies. He gave them the country where their campfires burned, and it would belong to their children forever. The Comanches had ridden into Texas on horses, bringing with them knives made of buffalo ribs, clubs, bows, and arrows headed with flint and obsidian. For a long time, there had been peace. For a long time, the tribe had been prosperous.

But now their faces were etched with sadness.

And anger.

And hatred.

The Comanches had no home. They neither inhabited villages, herded cattle, nor planted corn, but chose instead to follow the unpredictable life of a nomad, always on the move, seldom out of the saddle. Children had mastered the art of riding by the time they were three years old. It was a way of life.

In camp, the Indians' tents were fashioned from dressed buffalo skins, and the Comanches seldom remained in one place for more than three days, as deer and antelope and buffalo led them on. The search was unending. Sometimes, when drought spoiled the land and burned the grasses dry, they found little to eat at all: a few roots, prickly pear, and wild fruits of the season.

Assistant Surgeon Ebenezer Swift recalled,

> A party arrived at Fort Martin Scott, on their way to San Antonio ... Col. S., in command of the post, had orders to issue rations only to chiefs and their families, which was wholly inadequate to their necessities, having fasted three days. Two horses, which had died several days before in the stables of the garrison, were found in a ravine, and, cutting off the flesh (in a state of decomposition) from the bones, they prepared it for food for supper and breakfast, and proceeded on their way.

The white man, they said, had chased their game away.

They starved.

And the white man had nothing edible to give them as they died upon a land that the Great Spirit had given them.

Ebenezer Swift also noted: "[The Indians] are learning to drink whisky; many drink it now, who would not three years ago."

Resentment boiled within them.

Whiskey would pull the trigger.

To the Comanche, thunder was the voice of a big eagle, and lightning was his breath that changed the wind and made it rain. It was the source of the tempest and the tornado. Gunfire was the thunder. A tempest rose up from the smoke it left in a pungent mist across the prairies.

It could not be calmed.

It must be defeated.

The war had begun.

Colonel Albert Sidney Johnston of the Second Cavalry wrote of one Indian party:

> I regret to say [it] succeeded in surprising a party of seven citizens of that region and in killing two and wounding three of them.
>
> All the troops at my disposition have been kept actively and vigorously employed in repelling the attacks of Indians on the frontier and with comparative success. Yet, when it is considered that the frontier embraces an extent of more than a thousand miles, . . . it will be perceived that the means, to cover defensively such an extent of country, are inadequate and it may very well be doubted whether a large force . . . can prevent the incursions of small parties . . . unless they are followed to their hunting grounds with a sufficient force to act vigorously against them, they will continue their depredations.

Johnston's were not optimistic observations.

But they were correct ones.

At Fort Clark, Captain James Oakes of the Second Cavalry dispatched the following message:

> Unofficial but reliable information has been received that the mail party on its way down was attacked by Indians, and most of it destroyed. Use your discretion as to the size of the escort to be furnished from your post for the express, desiring it may be sufficiently strong to make itself respected, and, if the strength and condition of your command will admit, strong enough to attack and punish any party they are likely to encounter.

The army was preparing for battle.

Patience ran thin.

Then it ran out.

Adjutant General D. C. Buell, from the U.S. Army headquarters in San Antonio, wrote to his western chain of forts:

> All Indians in Texas that are not settled in the Reservation under the control and protection of the government . . . are to be

regarded as hostile and are to be pursued and attacked wherever they can be found.

But that was the problem.

The Indians, it seemed, were at times almost invisible upon a broad, open prairie that had few places to hide them. They were masters of guerilla warfare, with an uncanny ability to strike quickly and vanish into the arroyos, leaving few traces and a confusing trail that dared anyone to follow. They had become an impossible foe, described by Albert J. Myer as "the fiercest savage warriors that the Army had ever encountered."

He wrote a friend,

> . . . you ride everyday and sleep every night ready to be called into action at a moment's notice . . . You know I have a habit of reading in my Bible at night. I could not think that night . . . on the march how singularly it lay at the head of my bed while by its side protruded the handle of my revolver, and a sword and loaded rifle were ready for my grasp. Day and night, in this country, one must be armed, even here in my quarters my pistol lies in its case loaded and capped for I cannot tell when I may be called upon to use it.

Those who trekked across Texas in the 1850s never really knew what danger surrounded them, or if adversity even followed the ruts that their wagons had left behind in the sunbaked soil. Still they moved westward, dragging civilization after them, tempting a fate that had not always been kind to them. Life was precious but cheap, easily lost, usually forgotten. No one trusted the shadows. Nerves frayed and sometimes unraveled. At night, they knew that death, perhaps, lay only a silent footstep away.

Albert J. Myer wrote on December 2, 1854:

> This is quite an exciting night. The wolves are howling around the camp as I write to you and this evening we found near us the trail of hostile Indians. They may be near us now. But a Sentry is pacing before my tent and, we three, two officers beside myself have our tents pitched side by side, our revolvers are loaded and ready and I suppose we shall sleep as you will — I sprang up rather suddenly as I wrote that word for three rifle shots were fired from the wood near my tent and I heard my sentry bring his musket from his shoulder. He is challenging now 'who comes there?' and a voice answers 'Friend!' so it is at least a well educated Indian if any! I believe I will go to bed now. This little alarm will make my sentry careful and he will watch well.

The troops at Fort Clark rested beneath the shade of live oak and pecan trees. The mounted patrols that rode away from Las Moras Creek scoured the countryside with diligence and mobility that scattered them far and wide across an angry prairie. But then, horses were not always a luxury afforded the troops. Myer mentioned in one letter:

> I hear that an expedition against the Comanche Indians is on foot, very probably I shall be ordered to join it . . . There was a stampede on the road (toward Ft. Clark) . . . Indians carried off sixty-five mules and left the wagons in the midst of the prairie! Pleasant very, for the drivers! A day or two after they attacked a small party and carried off all their animals. This looks like work and I shall travel with my eyes open. We can whip them in a fair fight . . . I shall not be surprised if we meet the gentlemen but we shall assure them in the most polite manner that we do not desire their company. The way in which this is done upon the prairie is by pointing a rifle and occasionally by firing it.

The cavalry stayed in the saddle, sometimes fighting as many as three skirmishes a day. They searched, and they pursued their prey across a land that seemed to stretch out before them without end.

During one expedition, the cavalry dogmatically chased a band of retreating Indians for more than 600 miles, finally reining to a halt only after the braves had arrogantly crossed the Rio Grande and disappeared into the protected refuge of Mexico.

The life of the cavalryman was a frustrating one, spent — and usually wasted — in hopes of catching an enemy as elusive as the whirling dust devils that sudden windstorms would kick loose upon a desert floor. They were fighting a war that no one could win. They were the hunter. At times, they were the prey. They rode in triumph. They died upon the face of an unfamiliar land. They dutifully set up ambushes, waiting for the Indians to return from the shelter of Mexico. More often, they charged blindly into traps themselves.

The cavalry found a measure of revenge during the Devil's River Campaign of 1859. The military had learned that a large unruly band of Lipans had begun congregating just below Dolan Falls, raiding settlers who had pushed beyond the western fringe of civilization, running away with both horses and cattle. Pioneers were left homeless. They were left stranded. A few were even left alive.

The raids had to be stopped.

Fort Clark was immediately reinforced by additional columns of cavalry and dragoons, and an expedition rode away to meet the Lipans on uncommon ground. The soldiers were merely strangers in a strange land. The Indians waited.

In May, the Lipans ambushed a cavalry detachment about eight miles north of the mouth of Devil's River. The troops dug in along the canyon walls, pinned down and cut off from the unit. The soldiers were vastly outnumbered but not out-fought. The battle wore on with the day, and slowly the Indian line began to bend, then crack, and amidst the turmoil and the screaming and the gunfire, some broke and ran. For a time, the Lipan stronghold on the western shoulder of Texas was weakened, and settlers unwisely began striking claims further west and further beyond the strained reach of the military's protection.

When Colonel J. K. F. Mansfield inspected Fort Clark, he reported, "Indians are not known here except as marauders, who infest the road from time to time, & come from a distance. They have no abiding place in this vicinity & are in small parties."

By 1857, rock had been hauled up from a quarry for the construction of permanent buildings at Fort Clark. Mansfield was not impressed. The troops at the post, he believed, were obviously ill trained for the mission that lay ahead of them. They faced difficult circumstances, made even more troublesome by the lack of good equipment as they marched into battle. He wrote of his findings:

> Aggregate force at Command: 6 Officers, 240 men, 1 Asst Surgeon, 1 Ordnance Sergeant. In addition 3 men temporarily at the post.
>
> Of this force a large number were recruits & those of the two rifle companies on parade, were mostly uninstructed. The old soldiers of these companies were out on a scout after Indians that had recently attacked Major Hill Paymaster on the Devil's River; and as escort to Inspector General. These Companies therefore did not drill nor fire at a Target. They were neat and good looking men.
>
> The Artillery was also neat & well formed, but with so many uninstructed recruits, as to show but an indifferent Company drill, which was conducted by Col. Magruder, and afterwards by Lt. Eddy. They had not been instructed as skirmishers nor at the bayonet exercise.
>
> Company H. mounted rifles is provided with 2 buglers & 2 farriers & smiths & well supplied with recruits. The men are quartered in miserable hackales, no iron bedsteads. The horses are provided with a good corral & under large shade trees; but many of them worthless for this service . . . I condemned 13 horses to be turned into the quartermaster as soon as practicable & one to be dropped as having been foundered & gave out the 1st day out from Fort Clark & by my orders left on the road.
>
> Guard House is of stone with a good prison attached with 3 cells for solitary confinement, 1 general cell & 1 prison room — a

guard of 9 men — There were 12 prisoners, 4 deserters, 7 drunkness & one under sentence for stabbing another.

Colonel Mansfield wrote of the hospital:

> At present it is a log building and hospital tent. But a new and ample stone building with new stone kitchen & dead house & shingled roofs are almost finished for use. This hospital will be ample & very important, as on the route to El Paso from both Forts Duncan & San Antonio. And in case of distress & sickness to citizens travelling of good benefit to them also.

He said of the magazine:

> The walls of this building are up & of stone, but as there was an order not to use building materials without orders, it was suspended & no roof on it. The ammunition of the post is kept therefore in the commissary store & in good order . . .
>
> I would recommend that stone quarters for the officers of these companies be immediately commenced here, which would not cost much aided by the soldiers on extra duty . . .
>
> There appeared to be great harmony among the officers & a mess was established for the unmarried officers.

The command at Fort Clark, in 1857, asked for musketoons but was told that the weapons were not deemed "a useful arm for cavalry." And General David E. Twiggs ventured west with new ideas about how to conquer the Indians who, settlers warned, would forever stand in the way of progress. He didn't want war. He wanted to pursue peace, believing that his troops were fighting a useless conflict anyway. Twiggs wrote:

> Posts are so distant from each other that marauding armies of Indians can easily pass between them without being discovered; and, if discovered, it is very difficult to overtake them; indeed, there is not one case in fifty where a command can come up with them. The posts are situated at the most eligible points for the protection of the frontier, but 10 times the number of posts and men we now have cannot give entire security to the inhabitants and their property.
>
> I would respectfully suggest a change of policy. All Indians, for the past two years, found in or near the borders of Texas, have been treated as hostile. Would not a treaty or talk with them . . . lead to some good results?
>
> . . . Carrying on a war like the present is exceedingly annoying and harassing to the troops.

To the general, a treaty made sense. It seemed like a logical, viable alternative to bloodshed.

It didn't take General Twiggs long to change his mind. Two months later he sent out another letter which said:

> I have met with Mr. Neighbors, agent for the Texas Indians, and he thinks it is useless to have a treaty with the wild prairie Indians, for they have never observed any treaty stipulations and will violate all of their promises . . . I have seen and conversed with one of the Comanche Chiefs, and I am convinced his opinions of a treaty are correct.

So the troops daily patrolled the western terrain of Texas. Scouts hunted Indians and tracked down a few. They, like the Comanches, learned to hit and run, and more soldiers suffered from ague, catarrh, and pneumonia than from battle scars. As Assistant Surgeon Basil Norris had reported: Such sickness "has appeared in those men whose service in the field or on escort duty has compelled them, after much hardship and privation, to sleep upon the ground, exposed to rains and northers, which, although unfrequent, sometimes overtake them with all the rigor of the coldest climate."

Occasionally, large troop movements spread out across the plains. In the summer of 1856, for example, an entire Mounted Regiment was ordered from Fort Clark to New Mexico, further broadening the scope of military liability. The soldiers were to march a thousand miles, travelling slowly, taking as much time as needed to save the animals from being ruined and destroyed by a broiling July sun. Mrs. Lydia Spencer Lane rode in an ambulance — pulled by two gray horses — with her husband, a lieutenant, and she remembered the nights:

> . . . lights were out, and nothing could be heard but the tramp of the sentinel, the rattle of the chains by which the mules were fastened to the wagons, and the steady munching noise made by the animals while chewing their corn. Frequently the coyotes came outside the camp and serenaded us with their dreary, melancholy howls and barks, but we were too weary to be disturbed by them. All was peaceful, and it was hard to believe that behind that little rise, or clump of grass, Indians could easily watch what was going on, and be ready to run off any stray mule or horse that chanced to wander their way.

There were hardships:

> No one knows, who has not been deprived of these necessities, what a luxury a little milk or a pat of butter becomes when unobtainable, which was usually the case with us when travelling. Often in Texas, when we tried to buy milk at a ranch, where there

were thousands of cattle, there was not a drop to be had. The owners would not take the trouble to have it even for themselves.

There was fear:

I was constantly on the lookout for Indians, and a number of bayonet-plants together had given me many a scare, assuming in the distance almost any shape — men on horseback and on foot.

And there was misery to keep her company:

A severe storm of wind and rain set in the night, and by morning it was so cold . . . We had sheet-iron pans filled with hot coals and ashes put into the tent to heat it . . . all our blankets and shawls in the ambulance were more than damp . . .

The prospect was gloomy enough; we had nothing to eat with us, and the soldiers were hungry and wet to the skin. After watching and hoping against hope that the wagons would certainly come after a while, a man rode into camp with the information that they were ten miles behind and up to the hubs in mud! Pleasant prospect for such a night, — pouring rain and no provisions. We were in a grove of cotton-wood trees, and the men soon started a big fire. It was unnecessary to be cold, even if wet and hungry. Just at the darkest moment a train of wagons was heard approaching, and it proved to be one going down the country empty. The wagonmaster was able to supply the soldiers with rations for a meal, and we gladly accepted some bread, bacon, and coffee from their store, and felt wonderfully cheered after a hot supper.

The western migration continued to follow the pathways that the soldiers from those lonely outposts had cut deep into the prairie lands. Settlers nailed down homes, then clustered together and built towns, almost always within the shadow and the judgment of a military fort.

Just outside of Fort Clark, the little village now known as Brackettville prospered, nothing more than a rendezvous place for traders and trappers and scouts. In its streets, soldiers quenched their thirsts, traded insults, and took out their frustrations on each other. They fought to win, celebrated when they did, then fought some more. And the army placed metal cages around town and in back alleys to hold its unruly troops — keeping them off the streets and out of further trouble — until they could be hauled with aching heads back to their duty posts again.

The town was soiled but lively.

It had few morals but plenty of money.

Some damned it, and some worshipped it. It all depended on their motives at the time. Brackettville did offer relief from boredom and from riding too many miles for too many days upon a terrain that had neither mercy nor compassion.

One officer described Brackettville as "the ulcer of [a] garrison, an inevitable fungus growth — profusely plastered with mud, used for whiskey shops, gambling saloons." Progress did indeed have its problems, but few that temperance and a little sobriety couldn't overcome.

Then into town rolled the wagons of Henry Lawrence Kinney. He was a maverick, a soldier of fortune who had fought with Abraham Lincoln, Zachary Taylor, and Albert Sidney Johnston in the Black Hawk War up in Illinois, then built the first building in Peru, Illinois. He traded out of Havana, Cuba, got himself mixed up in Florida's Seminole war, dreamed of settling a colony in Nicaragua, and got the idea of hauling prospectors west to the gold fields of California. He advertised, as far away as Europe, that "if you want to travel west, I will supply the wagons and offer safe passage across Texas." It was a promise that he usually kept, especially if the gold seekers had enough money for him to hire a mounted guard. For Henry Lawrence Kinney, the county ruled by the guns of Fort Clark, was named.

Through Brackettville ran the stageline that Jems Birch had established from San Antonio to San Diego, a trip that took, at best, thirty days. The cost from San Antonio to El Paso was $100, with a $200 ticket required to make the perilous journey all the way to the California coast. Birch guaranteed seating on one of his fifty Concorde stages for all but 180 miles of the excursion. Passengers had to mount up on mules to traverse the wastelands of the Arizona desert.

But then, any pilgrimage across Texas could be unpredictable and sometimes cursed with danger. And the Indians weren't always to blame. On the day before Christmas, Mrs. Lydia Spencer Lane left Fort Clark for San Antonio, and she left behind a vivid description of the journey:

> The weather was like summer, and the evening was so warm in camp we were glad to get out of the tent for air. By morning a stiff norther was blowing, and water in a bucket in the tent froze to the bottom. It was bitter cold, and we were so anxious about the baby, fearing she might freeze to death. Our ambulance was better calculated for a summer ride than a journey on a freezing winter's day. Our driver, Biles by name, had begun very early in the morning to celebrate Christmas by taking a great deal more whiskey than was good for him, and which he procured from some unknown source. As it was a warm day when we left Fort Clark, he,

soldier-like, 'took no thought for the morrow,' and forgot his overcoat. We found out as soon as we started from camp that the man was too drunk to drive, and we had not gone far before he became unconscious. He was propped up on the front seat beside husband, who drove, and who occasionally administered a sharp crack over his head with the whip, to rouse him and keep him from freezing to death. I sat behind, with the baby on my lap, completely covered with blankets to protect her from the wind, and many an anxious peep I took to see how she fared, lest, while keeping her warm and excluding the cold air, I might smother her.

There we were, travelling over the prairie, far from any settlement, with no escort, and a young baby and a helplessly drunken soldier to be cared for. It was an anxious day for us, and we were much relieved when, late in the afternoon, we could see the little town of Dhanis in the distance, where we would find a fire and the assistance we needed.

The West beckoned for those tough enough to take it and hold it, and many were trying. Life wasn't easy, and for some it would get worse. Land was free and full of promise and, for a time, the Indians had been pushed aside and their buffalo scattered.

The Comanches brooded and hated, and they watched the coming of loaded wagons leave crooked ruts upon the land of their fathers.

Henry J. Veltmann, in 1859, was working in his general merchandise store in Dhanis when two cowboys walked in, and Alex Hoffman asked for some caps to put on his pistol.

"Why?" questioned Veltmann. "You've got caps on your cylinder. What's the matter?"

Hoffman shrugged. "I don't know why it is," the cowboy answered, "but it seems to me that we are going to have trouble on our trip hunting and I want to make sure to have my pistol in good trim."

He and Jeb Wolf smiled, checked their revolvers for the last time, and swaggered out of the store.

The next day, both were dead.

Henry Veltmann later recalled,

> A lot of cowboys got together and went to the place (west of Dhanis) and found both under a large tree where both had made a stand to defend themselves. The Indians had scalped Mr. Wolf and mutilated his body some, but Mr. Hoffman they didn't scalp him as he was somewhat bald headed.

Peace, to the Comanches, was humiliating.

They had not been beaten nor had their spirit been broken. Yet they knew that they were slowly losing the last strong grip they would have on an unsympathetic frontier.

And far to the East, the Indians saw seeds of war drifting indolently toward Texas. Perhaps, they thought, it might save them.

War was the only hope they had.

4

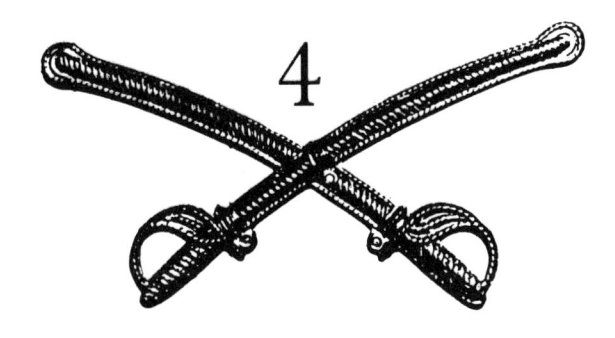

The Changing of the Flags

The South was restless.

It was holding fast to a proud way of life and on the verge of losing the traditions that had made it strong. There were rumors of open conflict in the air, and they spread toward the western frontier of Texas.

Lieutenant Colonel Robert E. Lee, in 1855, had formed the Second Cavalry regiment that one day would be known as the Fifth. A few years later, he had become a member of the Judge Advocate's staff, riding from Fort Mason to other outposts on the prairie of Texas and performing his court-martial duties. Many believe that he held court at Fort Clark, as well as Forts Lancaster, Inge, Duncan, Brown, and Ringgold. But no one knows for sure. As the U.S. Army points out, "There is no record that Col. Lee was ever at Fort Clark. There is also no record that he was not at Fort Clark either."

The threat of war above his homeland hung heavy around Robert E. Lee's shoulders. He, as a Union officer, had sworn to defend and protect the flag of the United States of America. Yet his allegiance lay with the South, the land of his birth, the land that held the graves, the dreams of his ancestors. On the muddy streets of San Antonio, General David E. Twiggs even tried to arrest Lee for being a Southern sympathizer. But the colonel snapped back, "Sir, I am still an officer in the U.S. Army and my sentiments are my own business."

Twiggs didn't trust Lee.

Robert E. Lee rode home to Virginia to clad himself in Confederate gray. The cries for secession echoed throughout the Southland, and Texas listened as the muffled sounds of war crept steadily closer to its own soil.

Sam Houston, the old Texas warhorse, the former president of the Republic and governor of the state, could not condone the tenets of slavery. It grieved him to watch as Texas tore itself loose from the spirit of the Union. Traveling the country, he said, time and again:

> Some of you laugh to scorn the idea of bloodshed as the result of secession. But let me tell you what is coming ... Your fathers and husbands, your sons and brothers, will be herded at the point of bayonet ... You may, after the sacrifice of countless millions of treasure and hundreds of thousands of lives, as a bare possibility win Southern independence ... but I doubt it.

Houston had made one fervent plea for Texas to remain loyal to the Union when a burly Brazos River bottom farmer elbowed his way to the front of the crowd and shouted, "But, Gen'l Sam, we can whip them damn yankees with cornstalks."

Houston smiled a sad smile and nodded. "Son," he said, "you're right. But them damn yankees won't agree to fight with cornstalks."

Off the coast of South Carolina, Fort Sumter shuddered amidst the thunder and the lightning of gunfire. A white flag was slowly raised above its stone ramparts, and a nation bowed its head beneath the burden of war.

In February of 1861, General David E. Twiggs officially surrendered eighteen Federal forts to the Texas Commission. On March 19, the Union troops marched away from Fort Clark, leaving the post on Las Moras Creek to a band of Texas volunteers under the command of Lieutenant Colonel John R. Baylor. The force was small, led by Captain T. T. Teel, and not particularly well equipped. The post's cannons were even dragged overland and placed among the coastal defense fortifications of Galveston. Settlers looked nervously out toward the prairies that hid the Comanches. The Indians eased closer again to the ranches.

The protection was gone. War lay at the doorstep of Fort Clark, but the enemy would be wearing warpaint, not Union blue.

John Salmon Ford, known as Rip, was named commander of the Rio Grande country and he realized that he must move quickly to prevent both Indians and Mexican bandits from conducting a "civil and fratricidal war" in the region. He met with General Guadalupe Garcia, assuring him that the Confederacy had no imperialistic demands at all on the land south of the border. Garcia, in turn, pledged his friendship and his support, and Ford immediately dispatched ten companies of the Second Texas Cavalry to Fort Clark, Fort McIntosh, Camp Wood, Fort Inge, Camp Stockton, Fort Lancaster, Fort Davis, and Fort Bliss.

The cavalry, on a far-reaching battlefield, would be his most efficient, most deadly weapon — a unit adapted from the European dragoons. It was comprised of soldiers who had been trained to fight on horseback or on foot, men who could serve as scouts, messengers, or a hard-charging strike force. Under Rip Ford, they formed daily patrols, assigned to chase down and disperse renegade Indians who had left the smouldering ruins of widespread, isolated homesteads in their wake. Fields were scorched, buildings burned, and women and children scalped, then left to die upon the sparse grasslands of an unforgiving prairie.

From across the border, Juan Nepomuceno Cortina and his merciless army of outlaws rode hard upon the small adobe villages that sprawled across the drought-stricken earth between Brownsville and Laredo. They left behind death and destruction and the hanging corpse of a county judge, a stern warning to those who would dare stand in defiance of Cortina.

Ford, trapped in the midst of three wars, prayed for reinforcements, and he received two infantry companies of Colonel Philip N. Luckett, as well as a howitzer battery. He walked out to review the new troops and discovered a haggard, unkempt band of hungry Mexicans and soldiers of fortune who would desert for "a few dollars and a little whiskey."

Men were scarce.

Good men were virtually impossible to find.

Supplies almost never reached Ford.

For months, the 800 horsemen of his Second Texas Cavalry were fed a starvation ration of dried beef and flour. Throughout the summer and on into the autumn, the men fought on empty stomachs, riding into battle with torn and ragged clothing. The glamour of war had escaped them. The words of Sam Houston came home to haunt them.

Many, whose homes had been sheltered on Texas soil, had agreed with Houston. They were morally opposed to secession and proud to have taken the oath of allegiance to the United States. Some had no desire to fight for the Confederacy. They had no desire to fight at all. The thought of slavery sickened them, and they fled, with sad faces and heavy hearts, from the tumult and the shouting of war.

A band of German Union sympathizers from Fredericksburg, Sisterdale, and Comfort pointed their wagons toward Mexico, fleeing those who had ordered them out of Texas. Some hoped they could live in peace on the far side of the Rio Grande until the sounds of battle had ceased, until they could return to their homes again. Others were planning to catch ships that would carry them to Northern ports where they could enlist in the Union Army. About mid-morning, on August 9, 1862, a group of sixty-five men, including John Sansom and Edward Degener, reined their horses to a halt beside the Old Dutch Waterhole on the west prong of the Nueces. They planned to camp for a couple of days, maybe even three, although the safety of Mexico sprawled only thirty miles away. The German settlers were in no hurry. War was definitely coming. But the gunfire and the dying lay a long way from them, and they rested beneath the hot summer sun.

That night, two guards walked the outer perimeter of the camp, mostly to make sure the horses didn't wander away. They didn't look for trouble. They didn't expect any. Gunshots in the early-morning darkness killed them both.

There are still differences of opinion as to whether the attackers were Southern renegades, (not in uniform) or Confederate soldiers. What is definite is that a band of ninety-five men, commanded by Major James M. Duff and Lieutenant C. D. McRae, swarmed into the sleeping encampment. In the darkness, the German settlers didn't know if the raiders were soldiers, or maybe deserters, or maybe political fanatics who had sworn to punish anyone who refused to remain loyal to the Stars and Bars of the Confederacy. The Battle of Nueces was swift and decisive. Nineteen Germans lay dead beside the Old Dutch Waterhole, and nine had been wounded.

Lieutenant McRae, the Southern commander, looked over the silent battlefield and found that two of his own men had been slain. Another eighteen had fallen. McRae hastily sent word to Fort Clark, almost twenty miles away, asking for medical assistance. The German Union sympathizers, without ceremony, were tossed into a common grave, and McRae rode out across the terrain in search of the Germans who had escaped in the confusion of the fighting. He found eight hiding nearby and killed them. Two months later, another seven would be shot down as they tried to wade and swim their way across the Rio Grande.

Back beside the waterhole, Lieutenant Luck calmly told his

men to turn each of the wounded German prisoners onto his stomach. And one by one, the lieutenant coldly executed all nine of them with a pistol, then ordered them stripped of their clothes and boots. It would be three years before their relatives came to gather up the bleached bones and carry them back home again to the hill country soil of Texas.

On August 12, the southern wounded were loaded onto crude litters, fed prickly pear apples, and dragged back across the bloodied plains toward Fort Clark. An army doctor and ambulance met them a few miles from the outpost, and the men spent their days at Fort Clark until the wounds had healed.

Only John Sansom and nineteen other German immigrants escaped the massacre at the Old Dutch Waterhole, running and hiding until they, at last, had the chance to join the Union Army.

Back in San Antonio, during the winter of 1863, Ford began recruiting men for "The Cavalry of the West." It promised glory. It only bound men to a three-month tour of duty. Rip Ford's voice could be heard saying, "Come to the front and aid in expelling the enemy from the soil of Texas."

By February, the Cavalry of the West had swollen to a thousand men, and Ford felt sure he could find another thousand soldiers as well. He appealed to them "in the name of patriotism, of liberty, and all that is dear to man." He offered them fifty dollars apiece to "defend their homes & property, their wives and their little ones against the brutal assaults of an enemy who respects neither age, sex, or condition, who plunder . . . the homestead."

Many enlisted.

Others pocketed their fifty dollars, promptly got tired of the cold, the discomfort, and the hunger, and deserted. Some would fight. Some ran. When March rolled into San Antonio, Ford herded his 1,300 men together and said that with the help of God he would drive the Yankees into the Gulf of Mexico.

The frontier had long been forgotten.

The Confederacy reigned, and the military, perhaps in desperation, had turned its full attention to the coming invasion of Federal troops.

Out in Fort Clark, the post surgeon and a hospital steward had stayed behind when the U.S. Army marched away. Troops marched into and out of Fort Clark with regularity, some looking for battle, others fleeing it. A few were merely looking for a fight, any kind, anywhere.

Ben Thompson, with the unit commanded by Colonel John R. Baylor, stopped at Fort Clark before continuing his journey on to

New Mexico. He had one bad habit. He was always late to mess, and he was always hungry. The sergeant had no mercy on him at all, didn't really care whether Ben Thompson starved or not.

Late one evening, the sergeant saw him coming and announced loudly that he had no more bacon or candles for him. The rations had already been handed out and were all gone. Ben Thompson could just suffer. Thompson frowned. He knew the sergeant was lying. Time and again he had seen the grizzled, ill-tempered old army veteran load up with all the food that had been left over at the end of the day. He hoarded it. And Ben Thompson had no use, no patience at all for a selfish, greedy man.

As he left, he spotted a stack of bacon on a table near the door. "Who does it belong to?" he asked one of the cooks.

The cook glanced nervously over his shoulder at the sergeant, then whispered, "The laundress."

Thompson leaned against the door for awhile and waited on the laundress. She never came. So he took the bacon, jammed it into his coat pocket and walked away. Ben Thompson would eat that night after all.

Within fifteen minutes, the old commissary sergeant came running out into the yard at Fort Clark, shouting, "What damned thief stole the rations of the laundress?"

Thompson calmly replied, "I took 'em, but I didn't steal 'em. You can replace them out of the over-issue of rations you made to your own mess."

The sergeant turned and angrily snapped, "You did steal them, and you have to surrender them right now. And as to my making an over-issue of bacon and candles to my own mess, that's a lie."

Slowly he began walking toward Thompson.

There were fire and hatred in his eyes.

But Thompson's words jolted him. "Don't come any nearer, Sergeant," Thompson said. "If you do, you will repent it."

The sergeant kept coming.

No two-bit recruit could stop him.

He reached for his pistol, and Thompson fired. The bullet tore through the sergeant's body, then slammed into the legs of another soldier who was standing much too close to the argument.

Both men fell.

Ben Thompson stared at the crumpled, bleeding sergeant with neither compassion nor remorse in his eyes. Behind him, Lieutenant Hagler drew his sword and moved forward. "You murderer," he yelled, "I will cut you in two."

There was, Thompson saw, a touch of the devil in the lieuten-

The Changing of the Flags

ant's eyes. The officer was far too angry to listen to reason. He only wanted to avenge his fallen sergeant, and Ben Thompson, as ready to strike as a rattlesnake, knew that he had no chances left to take.

He slowly raised his pistol and fired again. The lieutenant staggered, grabbed for his neck, and pitched backward onto the ground. Thompson stood alone, clutching a smoking gun. He didn't run. There was no place to go.

A captain moved quickly toward him, ordering Thompson to surrender. He nodded, almost nonchalantly, and handed his weapon peacefully to the superior officer.

"I'll surrender to you," he told the captain, "and I would have yielded to either of the others had they sought to arrest me instead of kill me. I am but a private soldier; still, my life is as dear to me as that of the highest officer to him, and a good deal dearer, Captain Hamner, judging by the way these gentlemen threw theirs away."

Ben Thompson spent that night in the guardhouse of Fort Clark. He knew he was destined to spend many more there, even though the sergeant gradually recovered from his wound. For a time, Thompson thought that the lieutenant, too, would survive the gunfight. He lingered for a month, then for six weeks, and finally lost his battle for life.

Thompson would pay, the army swore.

He would suffer for his crimes.

The guards chained him to the floor, and Ben Thompson lay flat on his back as one trial was delayed, then another postponed. The pain, the misery became virtually unbearable, and the rough flooring tore the skin from his back. Thompson, as he lay staring into the darkness, gave himself two choices.

He would be a free man.

Or he would die.

At the moment, he didn't care which particular fate awaited him.

The guards no longer paid much attention to Ben Thompson. He wasn't going anywhere. And the imprisoned man quietly convinced a friend to bring him a box of matches. Thompson twisted and turned and finally managed to strike the matches and set fire to an old chair. The blaze spread rapidly, and a suffocating smoke boiled out of the cell. For a moment, Thompson was convinced that he would either choke to death or be burned alive. But the friend fought his way through the flames, freed Thompson from his chains, and dragged him to safety.

The prisoner had hoped to escape.

But freedom eluded him.

An ironic twist of fate, however, did not desert him.

His friend, within a week, was stricken with chicken pox, but the soldiers of Fort Clark feared that he lay dying with smallpox, a disease that was more deadly on the frontier than an Indian's lance. No one would get close to him, and no one would take care of him. The troops realized that a smallpox epidemic could literally wipe out Fort Clark. There would be more funerals than soldiers left to dig graves in the dry earth.

Ben Thompson volunteered to do what he could to ease the suffering of his friend. After all, the man had saved his life in the fire. It was a debt that could never be repaid, but Thompson would try.

The army left the two men alone.

Anytime, a soldier, from afar off, would yell, "How's he doing?" Thompson would answer back, "He's worse. I doubt if he makes it." The military circle around them widened, then vanished. Let Ben Thompson expose himself to smallpox and die, the guards reasoned. It would simply save them a lot of grief and headaches and the trouble of a trial.

Within three days, the chicken pox had dried up.

The friend was ready to ride.

Ben Thompson, in the dead of night, slipped down to the stable and began saddling horses. Both men fled the outpost on Las Moras Creek, traveling to another fort two hundred miles away where they served dutifully until they were at last discharged from the army.

Communication was slow.

It never did catch up with them.

Most of the men probably thought that death had overtaken Thompson and his companion in the desert anyway, and none seemed to worry about his escape from military justice.

As soon as Thompson had his discharge papers tucked away in his hip pocket, he rode again to Fort Clark and once more enlisted for the war. He was never tried at all for killing the lieutenant over a stack of bacon and a few candles. Since his original term of enlistment had expired, the army decided that its court martial no longer had any jurisdiction in the case. After the war, Ben Thompson, whose past was as black as the clubs and spades in his poker hands, became sheriff of San Antonio.

For a time, during the conflict between the states, Fort Clark had a purpose. Cavalrymen scoured the countryside, a proud protective force that kept the Indians far beyond the front porch of the last settlement west. But supplies were cut short, then exhausted. Rations were sent to other battlefronts, and the men suffered, then

The Changing of the Flags

lost heart. Finally the troops were withdrawn from the outpost and scattered elsewhere.

The stone quarters were abandoned.

There were no voices at all to break the silence of the live oaks around Las Moras Creek.

And the settlers faced the Comanches alone.

They weren't concerned with secession or slavery, blue or gray, right or wrong, Gettysburg or Bull Run. Theirs was a personal way of life. Their lives were at stake. And their lives were worthless, or so it seemed. The military had deserted them. They loaded their rifles and waited.

Death would come riding an Indian's horse.

Brackettville's Albert Schwandner, in his later years, would recall that day in 1864, when five Lipan warriors attacked him and his mother, tying him to a tree with suspenders:

> ... the horrible memory of mother being brutally murdered by those savage fiends has never left me. I saw them shoot arrows into her body, while with her last breath she implored me to escape. After mother was dead, they left me tied up while they went back and plundered the house ... When father came in that night, he knew immediately that something was wrong, as mother and I were not out to meet him ... Forgetting everything in his frenzied determination to track down our captors, he walked, or ran, to Uvalde, a distance of over forty miles that night ...
>
> When the Indians had plundered all they wanted to, we started out afoot and walked all day through brush and over rocky ground. When the Indians arrived in camp, far into the night, where they were joined by a large band, they stretched me out on the ground. My feet were bruised and swollen and I was tired and almost dead with fear.
>
> They had a big war dance. The chief led the dance, dressed in his moccasins, buckskin regalia with a crown of bright colored feathers hanging down his back ... I expected death at any moment, but my time had not yet come, as the dance ended by day, and I remember I was sick and worn out, but was surprised to be alive as I expected the same fate as Mothers's.
>
> The next day they tore the top of my garments off, and I suffered from cold, but at their command had to keep going. The band we met had horses, and they whipped me with a raw hide rope, because I couldn't catch the horses for them, as I was so small.
>
> I rode behind one of the warriors, and I remember we had only boiled meat (cattle or buffalo) without any salt or seasoning. The bread was made from prickly pear apples, mashed up and put on a rock to dry. I was very sick from eating the Indian diet ...

That night the posse of white men with my father caught up with us . . . The Indians wanted to get away with the stolen horses, etc., so they slipped away in the night to avoid the encounter, which they knew would take place in a 'fight to a finish' the next day.

During a brief skirmish, Albert Schwandner was close enough to his father to hear his voice, but he couldn't reach the posse. At night, two warriors always slept beside him, their arms pinned across him to keep the boy from running away.

He suffered.

He endured.

Albert Schwandner was only six years old.

The Lipans finally traded him to Pablo Ramos down in Querto Cienegas, Mexico, for whiskey and a horse. A year later, the boy's father finally found him, and he had to sell everything he owned in order to raise a ransom large enough to buy back his son's freedom again.

The strife had ended for Albert Schwandner.

It ended for the South as well.

At Appomattox, Robert E. Lee laid down his arms and the Confederacy surrendered. More than a month passed, and Rip Ford and his band of cavalry stumbled across 250 men of the Sixty-Second United States Colored Infantry in the gnarled thickets of Palmito Ranch in South Texas. The force had left Brazos Island in a blinding rainstorm, then —reinforced by another two hundred men — it came swarming across the brush country as officers shouted above the thunderous gunfire, urging their men forward to fight and, if necessary, die for the Union.

Ford, riding up and down the skirmish line of his troops, yelled, "Men, we have whipped the enemy in all our previous fights! We can do it again."

The Confederate cavalry surged forward, and the Federal unit bent badly, then cracked and crumbled, scattering throughout the thickets in a confused, chaotic retreat to Brazos Island. The war had ended. To die would be a needless sacrifice. But Ford was unaware of Appomattox at all. For seven miles, he chased the fleeing Yankee blue before reining his horse to a stop and saying, "Boys, we have done finely. We will let well enough alone and retire."

The gunfire faded.

War — at least between blue and gray — came to an end in Texas as well.

The rage of the plains Indians posed a much greater threat. For

The Changing of the Flags

too long, the West had been neglected, trapped in the stampede of renegades who would kill to take what they wanted and usually did. Settlers lay at their mercy when U.S. troops again returned to occupy Fort Clark on December 12, 1866. Captain John E. Wilcox, in command of Company C of the Fourth Cavalry, had been given the precipitous assignment of again bringing a measure of order to an Indian-ravaged frontier.

It would not be an easy task.

For a time, he would receive little, if any, support from the government that he represented with his life. The plight of the soldier on the western plains was, perhaps, best expressed by an editorial published in a military newspaper, written by a frustrated, harassed, and bitter trooper:

> The fact that this is a frontier does not seem to be known to the authorities at Washington or elsewhere. In 1867, when the blazing dwellings of the pioneers of Texas lighted up the sky from the Red River to the Rio Grande; when desolated homes, murdered women and captured children were everyday occurrences along our whole frontier, General Sheridan in a report stated that 'no Indian difficulties of any importance had occurred in his department; that the Red River was a sort of dead-line over which neither Indian nor Texan dared to cross, owing to the hostility of one to the other;' and, in fact, intimating that the Texas frontiersman was generally the aggressor — this, too, at a time when the garrison at Buffalo Springs was besieged for days by five hundred Indians, and when appeal after appeal had been sent to General Sheridan for arms and ammunition. On the plains, if a colored soldier is killed carrying the mail, telegrams are sent to the associated press, the great dailies of the country expatiate on the event, and the world is horrified over his death. But here, where the Fourth and Sixth Cavalry have been for four years, doing more scouting, more escort, more fighting, more arduous service than any other troops in the army, no credit is given, no one knows of their great services, and both officers and men 'waste their sweetness on the desert air.'

Rumors persisted that General Philip Sheridan didn't particularly like Texas anyway.

It was said that a native of Houston once told the general, "If there was only plenty of water and good society [in Texas], it would be equal to any part of the Union."

And Sheridan dryly answered, "Those were the only two things they lacked in hell, water and good society."

Fort Clark, during those tenuous reconstruction years following the War Between the States, obviously wasn't highly regarded or appreciated. By July 1, 1868, it was garrisoned by Companies G and M, Ninth U.S. Cavalry, and Companies C and F, Forty-first U.S. Infantry. Extensive remodeling on the post had begun, with the addition of new stone barracks, officers' quarters, a headquarters, and a wooden stable 200 feet long. The water supply was sufficient, though not abundant, during all seasons. Grazing was generally poor for the livestock, a major problem for a cavalry post whose actions depended on the strength and the stamina of its horses.

When the war came to an end, the U.S. Army found itself with almost five million Yankee uniforms, and the government began shipping them to forts throughout the country, even to Fort Clark. The uniforms were hot and woolen and heavy, not suitable at all for the drought-stricken days in the Southwest Texas. By the 1870s, the troops at Clark would discard the uniforms altogether, guarding the Rio Grande in white shirts and Panama hats. However, a new commander, sent down from Washington, promptly ordered the vagabond soldiers to wear proper military uniform again, although he wrote his superiors asking for lightweight material that would bring his men a measure of relief in the Texas "tropics."

According to the *Descriptive Book of Texas*, published by the War Department, Brackettville was "a very unimportant place containing about two hundred inhabitants, principally Mexicans." Mail arrived three times a week. And, "an old Indian trail runs through this post to Fort Terrett."

The records indicated that the Indians who ventured into the area were

> Kickapoo, Lipan, and Comanche (unfriendly) . . . The Kickapoo, and Lipan Indians (professedly friendly) live principally in Mexico near the Santa Rosa Mountains, but they make occasional incursions hereabout, and kill and steal as much as possible. The Comanches infest the Pecos Valley, but occasionally come within a few miles of this Post on stealing expeditions."

The soldiers who had been commanded to protect the far western face of the frontier shouldered a burden that was heavy laden with misery and, more often than not, despair. The Indians wanted to kill them. Loneliness was an enemy, was just as lethal. Isolation sometimes became a torture of the soul. The troops often felt abandoned and stranded on the soiled edge of purgatory.

One trooper wrote of life that surrounded that chain of Texas forts:

The Changing of the Flags

> Situated on the outskirts of every military post may be seen a collection of huts, old tents, picket houses, and 'dugouts,' an air of squalor and dirt pervading the locality, and troops of shock-headed children and slovenly looking females of various colors completing the picture. These are the quarters of the married soldiers and of the laundresses, known in army parlance as 'Sudsville.' Each troop of cavalry was allowed four laundresses, who were rationed, and did the washing of the men at a fixed price . . .
>
> The 'officers' line' and their families always form the opposite side of the garrison from the troops, and as the subjects of interest in an isolated camp are comparatively few, and human nature in or out of the army is the same, a military post comes to resemble a little country village; gossip and scandal are rife, and in the vitiated atmosphere of the army, one so disposed need never be at a loss to hear or tell some new thing. This condition of army society is owing partly to its make-up and partly to the large amount of unoccupied and idle time which hangs heavily on its hands, and at which, as Dr. Watts says,
>
> 'Satan finds some mischief, still
> For idle hands to do.'

At Fort Clark, those idle hands — as well as the cheap whiskey and cocaine beyond Las Moras at Brackettville — kept many of the soldiers behind bars. Punishment was swift. It was certain. Sometimes, it was quite unusual.

One trooper crawled back from the sin-stained gutters of Brackettville one night and was promptly ordered to stand on the head of a barrel from reveille to retreat, taking only thirty minutes a day for meals. One sign was plastered to his back, and another had been pinned on his chest. On both were crudely printed the word, "Drunkard." Another soldier was forced to exhibit his wickedness while perched upon a rock. The common punishment of the fort was to carry around a twenty-four- to sixty-pound log from daylight to dark. It took all the joy out of carrying around a whiskey bottle from dark to daylight.

Chaplain Barr always prepared his church services.

But no soldier came.

Church attendance wasn't compulsory at all.

After months of disappointment, the chaplain was at last ordered to parade ground at 2 P.M. on Sunday while the garrison stood at attention before him. He was told to read the Articles of War. Then, once he had completed his assigned military duty, Chaplain Barr was allowed to conduct any kind of service he chose for the afternoon. None of the men ever left. They couldn't, not while still standing at attention.

He preached.

The troops listened.

No one escaped the condemnation of hellfire and brimstone after all.

Chaplain Barr would later report:

> In addition to the Divine Service for the troops, I have held a separate Divine Service every Lord's day for the families of the Post and neighborhood. These services have been regularly maintained, sometimes in the open air, in a store house, in hospital tents, and more frequently, in places less convenient . . .
>
> Weekly Services, Good Friday, and Services introducing the duties of a Post Military School, three times a week. Holy Communion administered four times. Baptisms, children 7; Sunday School children, white 29, Mexicans, 17 — in all 46. I superintend and instruct the School every Sunday morning.

The soldiers had little money to squander, and the high prices of the day took it all. Tomatoes sold for fifty cents a can, a glass of jelly would bring $1.25, and the cost of whiskey varied from $4 to $7 a gallon. Men did without tomatoes and jelly. Whiskey, in spite of the good Chaplain's warnings, became a staple. Cocaine, unfortunately, was legal and easy to get.

In May of 1879, Brevet Lieutenant Colonel H. C. Corbin reported that Fort Clark had quarters for 200 men, built of stone and in good condition. Four of the officers' houses had been made of stone and were quite livable. But three were constructed of "pickets with thatched roofs," and they were wretched places to call home. The storehouse occupied two stories, one for the quartermaster and commissary and one for the sales room. An attached stone building served as a granary — with a 3,000 bushel capacity — and a carpenter's shop. Corbin also pointed out that the fort generally kept three months' worth of supplies on hand at all times because the nearest subsistence depot was at the far end of a 126-mile wagon road in San Antonio.

That same year, Acting Assistant Surgeon Donald Jackson provided the following description of Fort Clark:

> There is no fort proper. The post is built in a quadrangle, one of whose sides, the northeast, runs nearly parallel to the Las Moras Creek, from which it is distant from 75 to 100 yards, and on an elevated ridge of nearly bare limestone rock, 40 or 50 feet above the level of the creek . . .
>
> In the general barrack room, which is 100 by 20 feet, there are two doors and two windows in front, and one double door and two

windows in the rear, with a large fireplace at each end. Ventilation is secured through the gable and ridge . . . There are no wash or bathrooms. Bedsteads are arranged in tiers, each 6 $^{3}/_{12}$ by 2 $^{10}/_{12}$ feet. There is a gun rack at one end and two shelves at the other, near the wall. These beds are placed at right angles to the wall, or across the barrack, in two rows. Bedsacks are filled with hay. Ordinarily each man has two blankets, of tolerably good quality.

The water closets are of a temporary nature, built of wood, situated about 150 yards from the barracks.

A kitchen and mess-room of stockade, and shingled, has been erected for one of the barracks . . . There is also used as a kitchen and mess-room for one company, a stockade grass-covered building, formerly used as a traders' store . . . Married soldiers and laundresses are provided with small tents . . .

Except the commanding officer, who has two rooms, no officer has at present more than one, exclusive of kitchen. The officers and men not provided with quarters occupy tents.

The privies are placed in the rear of the officers' quarters, at a distance of fifty yards, and are provided with portable boxes. There are no bath-rooms. Water is supplied from the spring by a wagon . . .

The guardhouse is a substantial stone building . . . and contains two rooms and four cells . . . The cells and prison-room are ventilated and lighted through grated holes high up in the wall, which are entirely too small, admitting an inadequate supply of air and light. There are no direct means of heating these cells or the general prisoners' room . . .

The hospital requires a new roof, new joists under the porch, a few doors, and other general repairs . . .

The post bakery is a substantial stone building . . . with shingled roof. In its rear is an oven capable of baking at one time 300 rations. There is neither chapel, laundry, nor schoolhouse at this post.

The stable was formerly on the opposite side of the creek in the rear of the commissary and quartermaster's store-house, being a stockade shingle-roofed shed. It is very much dilapidated . . . A new stable, 200 by 30 feet, of boards, with shingle roof, has been lately finished, situated 100 yards in the rear of the new barracks. It is divided into two rows of stalls . . .

The regimental library of the Twenty-fifth Infantry is kept at present by the regimental adjutant, boxed in his office. It contains 900 volumes. The post library at present consists of 184 volumes, kept in the quarters of the post treasurer . . .

The garden this season has been an entire failure, as the frost in the early part of the season destroyed the early vegetables. Scarcity of labor, as the men were busy working at the buildings, was another cause of failure. Fresh vegetables sell at almost fabulous prices, and, except onions, which are brought from Mexico, are very scarce.

Medical supplies are at present obtained from New Orleans, Louisiana, semi-annually.

Communication may be made with San Antonio, Texas, by stage twice a week, liable in the wet season to interruption from floods, which occur about half a dozen times a year. After leaving the post a letter usually reaches department headquarters in Austin in three to four days, and Washington from ten to twelve days.

The inhabitants of the surrounding country are principally Mexicans of the most worthless class. They subsist by performing the work required around the post, such as cutting wood and hay, and many of them have no visible means of support. There are a few industrious and energetic stock raisers in the country who thrive well, notwithstanding Indians and others who are constantly committing depredations on their herds.

For the women who lived at Fort Clark, life, at times, seemed to be unbearable. Some nearly starved. The butter was nothing but oil, and milk was thin, like water, with the flavor of wild garlic and onions. Too often, the beef was dry, and a strange, bewildering taste made it virtually unedible. Kitchens were without vegetables. And potatoes almost always arrived by wagon in a mass of decay. Women and men alike began to lose their strength and finally their health. And the army wives spent a great deal of their days wrapping wet cloths around their jars of milk and butter, trying to cool them, always wondering if they would be soured by suppertime.

They fought wasps and lost battles with roaches. One wrote,

> They not only covered the kitchen floor until it was black, but actually flew around our heads, and even invaded the bedrooms upstairs until life seemed intolerable. A thorough system of cleaning and scrubbing was instituted; for they love dirt, which was, in fact, the original cause of such an undue supply.

And, she complained,

> A picnic in Texas was simply impossible on account of the red bugs and wood-ticks, which were not only countless and disagreeable, but so poisonous that I knew an officer who had been obliged to camp out on the ground, suffer so severely from their attentions that hospital treatment was necessary for weeks. The sores caused by these insects are frequently very painful, because they bury themselves beneath the skin, and actually have to be dug out.

The vermin ran rampant, as it is wont to do. So did the scorpions and tarantulas and centipedes and snakes.

They were more of a nuisance than the Indians.

But they weren't as deadly.

Wives waited patiently, re-wetting the cloths, recooling the jars, as their husbands fanned out across the countryside when news of Indian raids finally found its way back to Fort Clark. The troops traveled out across the thick gray grass that clung to the prairie floor, beyond the live oak and pecan trees that stubbornly attached themselves to small stream beds and refused to let go.

Rattlesnakes crept beneath the shadows of limestone outcroppings.

And jackrabbits bounded madly toward deserts of sand and rock where the scorching summer sun was a misery to all but the cacti that defied it.

The Indians left a trail of blood and destruction for the cavalry to follow, then hid themselves arrogantly in the hills and canyons of Mexico. They didn't always escape. Acting Assistant Surgeon Donald Jackson reported that Captain Wilcox and the Fourth Cavalry had found a band of renegades on the Pecos River and abruptly ended their raids on mail parties. Jackson also wrote, "Major J. M. Brown, Ninth Cavalry, succeeded in completely routing the Lipans from the Lower Pecos."

Yet the troops from Fort Clark, too many times, saw their search for the marauding Indians stop upon the muddy, forbidding banks of the Rio Grande. The river, narrow as it was, served as a wall that kept them from invading a foreign nation, from chasing down an enemy that could not be pried out of its Mexican sanctuary. As Jackson pointed out,

> ... the Indians operating in this region have been the Lipan and Kickapoo tribes, whose homes are in Mexico. The military operations have been, therefore, comparatively futile, as but little injury can be done to Indians unless their homes are reached.

The soldiers waited and grew weary.

They were always on the move and always late.

Settlers lost their cattle and their patience.

And rumors pervaded Fort Clark about an impending war with Mexico. The wife of Captain Orsemus Bronson Boyd wrote:

> We were so near the border that whenever any marauding band of Indians or horse-thieves succeeded in capturing a herd of cattle from some neighboring ranch, they would cooly slip over the Rio Grande into Mexico with their booty; and by the time our troops, again and again called out, could overtake them, the marauders would have crossed the border, where capture was impos-

sible, because Mexico allowed no American forces to enter her territory without special permission.

Matters continued on that basis for years, infuriating our troops, who were delighted when it produced results that seemed likely to culminate in a war between the two countries.

But that never occurred, though its threatenings filled our post with troops until they formed a little army, which when mustered in full parade stretched in double columns across the immense parade ground, and made a beautiful sight; one which, seen daily, was so pleasing that we almost forgot the discomforts of life that surrounded us.

The soldiers didn't forget.

They were angry and annoyed and tired of being defeated by the waters of the Rio Grande. They rode hard and drank hard and cursed both the Indians and their luck.

They longed for a fight.

The commander of one scouting party, named Gaskill, led his force along 315 rugged miles in fourteen days, then filed the following report:

> Saturday 13th: . . . moving due east towards Nueces River and struck a large Indian trail after having traveled about ten miles. Trail led north west and was thought to be about four days old. Found three horses the Indians had abandoned one of which was so badly maimed and worn out that he was left; the others were taken along with the command. Immediately commenced pursuit traveling rapidly north east until sundown. Camped on West Fork of Nueces having made thirty six miles, a large portion of the distance over rocky and heavy ground. The Indians had camped one night on a mountain near the Nueces where the remains of a mule were found which had been roasted and nearly consumed . . .
>
> Saturday 14th: . . . Three pack mules broke from the column causing a delay of about two hours. Became entangled among hills and did not resume rapid march until about eleven A.M. Trail plain and showing that over one hundred animals were with the party. Picked up another abandoned poney . . .
>
> Monday 15th: . . . traveling rapidly along the trail and near the river . . . Trail there spread and was almost obliterated by cattle trails which caused a loss of two or more hours in searching the vicinity. Discovered a side trail of a party of six or eight Indians leading northeast which we followed over a very rough piece of country and then back over the mountains to the river . . . The main trail was found . . .
>
> Tuesday 16th: . . . Aroused the men at 3 A.M. and resumed

The Changing of the Flags

route at 5:30 A.M. The trail making almost due north along the dry canon . . . of the river. Saw one abandoned horse too much disabled to travel. At noon found rain water sufficient for the men and stock. Country open — the trail passing over ground apparently frequently traveled by Indians. At two P.M. came to a small pond of water fit for stock only. The grass was worn off all around with trails leading in all directions . . . At dark (having continued the march northward) went into camp without water for the men . . . No supper — the troops being supplied with flour which could not be used without water . . .

Wednesday 17th: Marched at 4:30 A.M. without breakfast or water . . . At six A.M. trail spread over a grassy plain causing a loss of time in foot trailing . . . At 9 A.M. came to a rock where a small amount of water was found, the first the men had had since noon of the previous day. The march was continued north west the trail entering a canon leading towards head of Devil's River . . . The trail led out of the canon and headed due north toward the plains. The command at this point had been without food since noon of the previous day and the guide assured me that there was no water within sixty miles in the direction the Indians had taken. They had passed in rear of rain which supplied them with water and furnished them to travel a route which we could not take . . . as two or three days of dry weather had completely dried up all rain pools . . . and it was thought best to abandon the trail and go south west in search of water . . .

Thursday 18th: . . . Made thirty miles by 5 P.M. and reached Beaver Lake . . . where the command camped. The pack mules were badly used up. One was hauled by main force about three miles to the water . . . No Indian signs.

Friday 19th: . . . marched along the river to Pecan Springs. Grass, water, and wood good.

Saturday 20th: . . . No Indian signs.

Sunday 21st: . . . met drover of cattle going west with escort of seven men from Fort Clark. Also learned that the Post Commander had sent a scout up the road to the mouth of the Pecos River to intercept the Indians we were following, a well-conceived idea had they been, as we first supposed, Mexican Indians, or rather those residing on the Rio Grande . . .

Monday 22nd: . . . Marched at six A.M. for Devil's River and San Felipe . . .

Tuesday 23rd: . . . Remained at San Felipe during the day to rest the stock.

Wednesday 24th: . . . arrived at Mud Creek at 10:30 A.M. Remained two hours and marched to Soldier's Creek and camped . . .

Thursday 25th: Marched at daybreak and reached Fort Clark at eight A.M. . . . The pursuit was as rapid as circumstances permitted.

At times the trail was indistinct and scattered that time was necessarily consumed in ascertaining its true course. I proposed to the guide that we make an effort to follow it during the night but he assured me (and I believe correctly) that the chances of going astray were great — hence the idea was abandoned. I am satisfied that the Indians belong to the Comanche tribe from the directions they took during the last two or three days to the point where I abandoned their trail.

The days had been long, the miles longer.

The men had ridden with dry throats upon a dry land that could not quench their thirst.

The Indians had not beaten them.

The terrain had.

They came home empty-handed, worn, and frustrated. They were engaged in a war they could not win, not, at least, until they found the enemy and left his refuge beyond the Rio Grande in ruins.

5

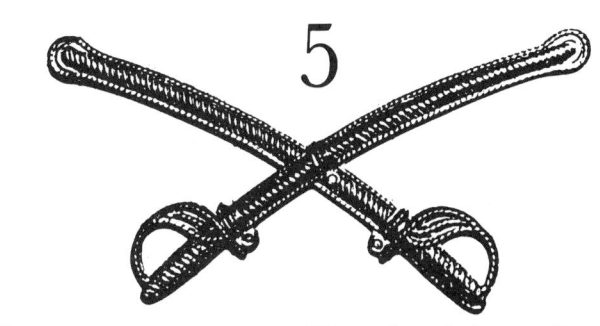

Invasion in a Forbidden Land

Captain D. A. Ward kept a close watch on the old Indian as he rode slowly out upon the Uvalde ranch, and he didn't trust him. John was different. He was peaceful enough, had not raised a hand to harm anyone along the Texas border. He did his chores and mingled with the cowboys who herded McMartin's cattle, and no one found anything at all bad to say against him. Yet John disturbed the captain from Fort Clark. Perhaps he was merely prejudiced against all Indians. Lord knows, they had caused him enough suffering and misery out beyond the Devil's River. Perhaps it was because he had discovered a large Indian trail, probably made by 200 to 300 head of mules and horses, angling northward from the ranch itself. Perhaps it was simply because old John never bothered to tell the same story twice that bothered Captain Ward so much.

He sat down and wrote the Post Adjutant at Fort Clark:

> There are circumstances that attach a strange suspicion to this Seminole Indian, which consist in the differing and contradictory statements regarding Indian property in his possession also regarding the name of the tribe to which they belong.
>
> His first statement was that they were Comanches, his next that they were Kickapoos and his 3rd that they were either Lipanes or Kickapoos. At first he said that he knew them and that they would not harm him. He claims that he aimed a gun at them

which however would not go off... This man should in my opinion be watched, not only during the full moon, but at all other times. This is a section of country that is advantageous to the going in and out of Indians.

It is impossible to determine positively whether they were Comanches or part Kickapoos and Apaches. An Apache chief's headdress was found by a man by the name of Cox (living on the lower Nueces) adjacent to the trail...

Lt. Davidson and party... reports no Indians in the country, but found the ranch men all confident of another raid ere long.

Maybe old John knew all about it. Captain Ward didn't know. He had fought the Comanche and Lipan and Kickapoo, was fully versed in their lifestyle and their warfare. But he didn't understand the Seminole at all. There were so few of them. And most were black. They came from deep in the mountainous bosom of Mexico, orphans in a land that regarded them as strangers.

At that moment, the captain would never have believed it. But those suspicious Seminole Negroes would one day provide the cunning and the daring that would at last free the Trans-Pecos from the deadly grip of the Plains Indians.

The Seminoles had been a dying, yet defiant race, at home in the tangled swamplands of Florida. At Payne's Landing, in 1832, a government official offered them $15,400 if they would move west beyond the Mississippi and leave their old land behind for speculators and settlers whose greed wanted to possess the eastern seaboard. Besides, he said, each man, woman, and child would be rewarded with a blanket and some sort of homespun clothing to wear. Their bellies would be full again, and they wouldn't have to walk naked out of the alligator bayous.

One young man suddenly arose and shouted: "The only treaty I will never make is this!" He slammed his knife into the wooden table, glared at the American officials, and whispered with malice in his voice, "I will make the white man red with blood. I will blacken him in the sun and the rain, where the wolf shall smell of his bones and the buzzard live upon his flesh."

Osceola had spoken.

And he led a band of Seminole fanatics on a lethal campaign of guerilla warfare against American forces, striking quickly and often, then vanishing into a mosquito-infested world where no one else wanted to go, yet everyone wanted to own for themselves.

Osceola had declared, *When I make up my mind, I act. If the hail rattles, let the flowers be crushed.*

Florida shuddered beneath the Indian's vengeance. He could

Invasion in a Forbidden Land

not forget that day when General Clinch had attacked his own village, and a witness would write:

> In an instant hundreds of lifeless bodies were stretched upon the plain, buried in sand and rubbish, or suspended by an explosion from the tops of the surrounding pines. Here lay an innocent babe, there a helpless mother; on the one side a sturdy warrior, on the other a bleeding squaw. Piles of bodies, large heaps of sand, broken guns . . . covered the site of the fort.

Osceola fought.

Yet his people hungered. The corn was gone, and they fed themselves with coontie and briars until they were gone, too. Their strength weakened, and so did their spirit. Some crawled to the army of General Thomas Sidney Jesup to surrender. Others begged for survival.

Robert Reid, the governor of Florida, looked out across the Everglades and said,

> We are waging a war with beasts of prey . . . We must fight fire with fire . . . If we drive him from hammock to hammock, from swamp to swamp, and penetrate the recesses where his women and children are hidden; if, in self defense, we show him as little mercy as he has shown to us, the anxiety and surprise produced by such operations will not, it is believed, fail to produce prosperous results . . . 'Lo! The poor white man!' is the ejaculation which all will utter who have witnessed the inhuman butchery of women and children and the massacres that have drenched the territory in blood.

Osceola, lured from the swampland by a white flag of truce, was captured and imprisoned. He had been betrayed. And his life was short.

Runaway Negroes had penetrated the mangrove swamps to live with the Seminoles and adopt their way of life. Both black man and red man were being hunted down like dogs, and they looked to Coacoochee, the Wildcat, as their final hope.

He had said:

> The land I was upon I loved; my body is made of its sands. The Great Spirit gave me legs to walk over it; hands to aid myself; eyes to see its ponds, rivers, forests, and game; then a head with which I think . . . Why cannot we live here in peace? I have said I am an enemy to the white man. I could live in peace with him, but they first steal our cattle and horses, cheat us, and take our lands. The white men are as thick as the leaves in the hammock.

For the Seminole, for the Negro who fought by his side, there would be no peace.

On the battered fringes of Okeechobee swamp, the Wildcat dueled with General Zachary Taylor and gave up the land he loved, the sands that had made his body. It had cost the U.S. government $20 million and the lives of 1,500 soldiers to remove only 3,000 Seminoles and runaway slaves.

The Wildcat had fought his last fight on Florida soil, and he and his people trudged slowly westward, burying their hopes and their dead along the treacherous Trail of Tears. Only their spirit would survive.

Sophia Dorothea Ruede, a missionary, had been a witness to their journey, and she wrote, "They presented a most striking appearance, more savage than I had seen. They were yelling and dancing and behaving awfully, so that my courage almost failed me to live among the Indians."

Their creed was a simple one: *Live as you please but die brave.*

By 1842, the Seminoles and blacks had been driven like sickly cattle and left on the poorest sections of the Creek lands in the Oklahoma territory, hungry and without water, cold and without clothing, broken and bewildered.

The Wildcat was an embittered man.

His people had been promised rifles and kettles and tools in the new land. There were none.

In anger, he wrote to General Worth:

> We have been conquered. Look at us! A distracted people, alone without a home, without annuities, destitute of provisions, and without a shelter for our women and children, strangers in a foreign land, dependent upon the mercy and tolerance of our red brethren the Cherokees; transported to a cold climate, naked, without game to hunt, or fields to plant, or huts to cover our poor little children; they are crying like wolves, hungry, cold and destitute.

The Wildcat had grown tired of the fighting and the dying. The Cherokees, living nearby, had offered his people help and sustenance. Yet, he had been given acreage among the Creek Indians, and he hated them. They were worse than the white man, stealing what little he had, waiting, perhaps, to even kill him. So the Wildcat walked away form the land he didn't want, heading to Texas and the Mexican border. In Eagle Pass, he told the Indian Commission in 1851, that "to avoid a war he had left that place and started to search for a new home."

Invasion in a Forbidden Land

Others, led by the Negro Chief John Horse, followed. The Seminoles, for so long, had allowed the blacks to be free men, living in separate villages, tending their own fields, and paying a tribute in corn to the chieftains who had accepted them like brothers into the foreboding swamps of the Seminoles. Through the years, many of them had intermarried with the tribe, and the two races became as one. And they were outcast.

For two decades, beyond the southern banks of the Rio Grande, their names and their faces vanished and were forgotten. They were gone, and no one missed them.

By 1870, the U.S. Army was searching for them again.

The Indians were making a mockery of the troops who patrolled that vast stretch of brush country that lay between Las Moras Creek and the Pecos River. They rode in small bands, moved as cautiously and as swiftly as the wind itself, and left behind scars of ruin but few tracks. The renegades struck ranches and settlements, and the cavalry would find only trails that were crooked and cold and led them nowhere. Men rode for miles and miles and found nothing but miles and miles. Mexico had swallowed up the Indians and kept them safe until they ventured back across the Rio Grande when their supplies were low or their anger at a fever pitch.

The army was convinced that it had to fight Indians with Indians if it ever hoped to win and free Texas from the Red Man's stranglehold. The cavalry from Fort Clark desperately needed scouts who could ride like an Indian, think like an Indian, and fight like an Indian.

Officers found a few Tonkawa and renegade Lipans who did not mind waging a war with their own people. But they could not be trusted. They betrayed their own. They would betray the cavalry just as easily, or desert when they were needed most, or get drunk and be unable to ride when they had a few extra dollars tucked away in their pockets.

The military remembered the Seminole Negroes who had pledged their allegiance to no one. They were tough and reliable and excellent fighters. The Seminole had always been portrayed as a warrior who was comfortable when all that he had was a blanket and enough fire to light his pipe. He didn't ask for much, and rumors persisted that the Seminole Negroes of John Horse had grown tired of Mexico, had become bitter enemies of the Comanche and Lipan tribes, and were wanting to return north of the Rio Grande again. Major Zenas R. Bliss of the Twenty-fifth U.S. Cavalry (colored) at Fort Duncan decided he would make it easy for them to come home. After all, the Seminoles and the Negroes had once been worthy foes. Perhaps, they would be worthy allies as well.

It was worth the gamble.

Major Bliss dispatched Captain Frank W. Perry into Coahuila, Mexico, and in Nacimiento, he found John Horse and his lost remnant of Seminole Negroes. Captain Perry carefully laid his proposition before them. If the men enlisted as scouts with the U.S. Army, they would be provided with arms, ammunition, and rations by the military. Each would receive the pay of a regular soldier. As soon as their enlistment period had expired, they would be given land grants in the states and allowed to remain as U.S. citizens.

It was more than anyone had ever offered them before. John Kibbitts, representing the Seminole Negroes, agreed to the treaty, one that the tribe would always refer to as "de treatment."

On August 16, 1879, Sergeant John Kibbitts, known as the Snake Warrior, and ten privates officially were knighted as cavalry scouts at Fort Duncan. During the next summer and autumn, Elijah Daniel brought another twenty scouts from Matamoros to Fort Clark. By the spring of 1873, another dozen recruits from John Horse's Laguna band had ridden forward to take advantage again of the military's promises. The years had tempered their anger, and the bitterness they had felt in the Florida swamplands had died within them. Besides, some of the scouts who pitched tents around Fort Clark were merely Texas blacks who had simply married Creek or Seminole women. Some were old-time soldiers who had been mustered out of colored regiments and who were merely trying to find a way back into the army again.

The scouts handed in their old-fashioned, powder-worn muzzle loaders and received Spencer carbines, then Sharps carbines. They would go to war in style, and they would be riding their own horses. The Seminole Negroes disdained military uniforms, preferring the garb of their Indian traditions. Major Bliss smiled and called them "excellent hunters and trailers, and brave scouts . . . splendid fighters."

Young officers were appalled.

They had been drilled in the spit and polish of West Point, and one of their reports about the Seminole Negro scouts said in disgust: "Discipline. Fair; Instruction, Progressive; Military Appearance, Very Poor; Arms, Spencer carbines — Good; Accountrements,, Good; Clothing, Fair." And a second report mentioned: "Clothing, Good enough for Indians." Those officers just couldn't believe that army-paid and army-enlisted scouts would be allowed to ride into battle wearing buffalo-horns or warbonnets.

Invasion in a Forbidden Land

Major Bliss wasn't concerned.

He was impressed by their tracking skills, their endurance, and their ability to fight, to stare at death eyeball to eyeball without ever blinking. They would definitely be, he believed, the weapon his military force needed so badly. Some spoke Spanish. Some spoke Indian. And some spoke both. Only those whose faces were wrinkled with age could still communicate with English. They looked toward the cold, crooked trails that the Lipan and the Comanche dared them to follow.

The young officers perhaps were not ready to accept the scouts in their ranks. But the cavalry who rode the prairies wanted any kind of help they could get. They knew the dangers. They had experienced the frustrations. They had seen the spoils of war, and sometimes it had sickened them.

As he marched several companies of the Ninth Cavalry from Fort Davis to Fort Clark, General Wesley Merritt heard the sharp staccato of gunfire in the distance as he approached Howard's Well. He promptly sent Lieutenant Vinson with a detail of men to investigate the ominous sounds, and the troops stumbled across an Indian band that had attacked the six freight wagons of Anastacio Gonzales.

Gonzales lay dead.

The wagons were ablaze.

Lieutenant Vinson spurred his horse forward and charged into the battle. It would be his last fight. Vinson fell dying on that harsh Texas soil, and his men quickly retreated to the command of General Merrit.

The Indians fled to Mexico.

No one could find them to avenge the young lieutenant's death.

Disappointment burned deep.

The cavalry could chase but could not catch the enemy it wanted most. As Captain Michael Cooney of the Ninth Cavalry reported on November 28, 1872:

> I marched from Fort Clark at 3 o'clock P.M.November 17 with my command consisting of 1st Lieut. Patrick Cusack and thirty enlisted men of Troop "A," 9th Cavalry, also a guide, with rations for seven days which was made to last ten days. I marched that night to Cope Ranch on West Fork of Nueces River to Kickapoo Springs with the intentions of crossing the country between the West Fork of the Nueces and Devils River. I found the country almost impractical for travel being alternate mountain and valley with neither high land or valley favorable to travel. No permanent water between the two rivers. Some water was found in niches

from recent rains. On the 21st we came in sight of Devils River but found great difficulty in getting down to it. However, after several hours search a place of descent was found and an Indian camp which appeared to have been abandoned about twenty four or thirty hours previous was found. I crossed over to the west bank and marched down the river in search of trail or their camp. After marching some distance a party of (4) Indians were seen on the east side of the river and coming toward the camp referred to as abandoned. They were driving eight animals. I detached Lieut. Cusack with a party toward them but they left the animals and rode off at speed. It was now dark and after search an ascent to the left bank could not be found. I drew in the detachment and encamped intending to cross as early as possible next morning. On the morning of the 22nd I sent Lieut. Cusack across as early as possible but the eight animals were in the same place and no trace of the Indians could be found. One of the animals had died during the night and of the seven remaining two died next night. We recrossed again to the west bank after resting the animals. The trail being found, meantime, it was taken at as fast pace as possible under the circumstances and followed to the El Paso road about two miles south of California Springs where it crossed. I found Lieut. Valois with his command on the trail at this point he having just arrived after marching out of Clark. We arranged that he should follow the trail immediately and I would encamp and follow in the morning. Lieut. Valois followed the trail as far as possible that night and was on it again at daylight. I was also on the march at daylight and pursued his trail. Lieut. Valois followed the trail to the Rio Grande River and could see signs of the Indians having crossed the day previous and encamped on the Mexican side that night. I met Lieut. Valois about ten miles this side of the Rio Grande and after him reporting these facts to me I gave up the pursuit and turned homeward . . . My animals suffered greatly from the roughness and want of grass. The men also suffered and were forced to lead and pull the horses over at least half the distance travelled but they did not complain and would feel compensated for all if they could only get a brush with the Indians.

Hatred for Mexico began to boil over among those ranchers and scattered settlers who sought to survive along the Rio Grande. They were convinced that the government south of the border was flagrantly guilty of harboring criminals in warbonnets who had circled their homes like vultures in the night. They didn't like it. In anger, several of them banded together, loaded their rifles, and attacked the little Mexican village of Resurrection.

Lieutenant Cusack marched immediately to the west bank of the river, hoping to quell the rebellion and calm the tattered nerves that had become unraveled along both sides of the Rio Grande. When he arrived, the gunfire had been silenced. Cusack signalled for the alcalde of the village to meet him at the riverbank. All he wanted was information about the raid. All he could promise was that justice would be carried out. The United States considered itself to be a friend to Mexico and wanted to stay that way.

But the alcalde cursed him.

And the alcalde blamed him for the attack.

On that day, there would be no friendship at all between the two countries. Such rebel uprisings could not be calmed. Goodwill soured. Relations were torn apart. The Ninth Cavalry found itself in conflict with the Indians, with the Mexicans, and with its own people.

General Augur reported:

> The labor and privations of troops in this Department are both severe. The cavalry particularly are constantly at work, and it is a kind of work too that disheartens, as there is very little to show for it. Yet their zeal is untiring, and if they do not always achieve success they always deserve it. I have never seen troops more constantly employed.

Their job was a demanding one.

It aged them long before their time, broke them, and left them to ride the prairies as hollow men, searching for a foe they could not find, hunting for Indians who were as elusive as the winds.

The Seminole Negro scouts would definitely make a difference.

While stationed at Fort Clark, they built huts for their families along Las Moras Creek, made from mud bricks held together by a little straw and stubble. The women tended the garden, and the men hunted or fished or fought. It's what they did best. Nearby, the scouts erected an adobe Seminole church with a thatched roof, about twenty feet long and eighteen feet wide. It had no pews at all, and the Seminole Negroes — all good Baptists — sat upon old chairs and boxes and logs. For communion, they drank cold tea, not wine, not grapejuice, because tea was more expensive, much harder to get. So it obviously was a much more important drink to be swallowed during the holy rites.

In May of 1873, the Seminole Negro scouts were placed under the command of John Lapham Bullis, a small wiry man whose face burned as red, sometimes, as the Indians he fought. He was a tireless man, and Frederick Remington would say that surely Bullis must pay high life insurance premiums because he seemed to have a knack for finding hostilities and waging war, especially when he was outnumbered, and he almost always was.

Typical was that September morning in 1871, when Lieutenant Bullis and four privates suddenly ran across three Indians driving away a herd of 300 cattle. General Order Number 17 of the Department of Texas reported:

> Lieutenant Bullis attacked at once and captured the herd. The Indians were pressed closely about a mile when they joined a party of fifteen more Indians and stopped to make a fight upon the top of a hill, being soon joined by ten more, making in all twenty-eight Indians. Lieutenant Bullis, with his four men, attacked them and maintained the fight for upwards of thirty minutes, but found it impossible to dislodge them and retired, taking with him, however, a second herd of cattle which the Indians had collected — numbering two hundred.

Bullis, always sporting a black mustache, and his Seminole Negro scouts were an ideal military combination on the West Texas frontier. Both traveled light and traveled fast and could live off the land without complaint. Sometimes they ate rattlesnakes, and sometimes they went hungry. But they never slowed down. And his scouts referred to Bullis as "The Whirlwind."

During the War Between the States, John Lapham Bullis joined the Union Army as a corporal with the New York Infantry and tasted firsthand the bitter conflict at Harper's Ferry. In July of 1863, he was taken prisoner at Gettysburg, then paroled, and promptly appointed as a captain in the 118th United States Colored Infantry Volunteers. Bullis was mustered out of service in 1866, and he tried for a long time to become a commercial businessman, supplying firewood to steamboats that churned up the Mississippi River. The venture was a risky one, more trouble sometimes than it was worth. A year later, Bullis returned to military service, enlisting as a second lieutenant in the Forty-first United States Infantry.

He asked to be sent West.

As commander of the Seminole Negro Scouts, John Lapham Bullis would become, some say, the greatest Indian fighter of them all. His men believed in him. He was tough. He was "The Thunderbolt." One of his scouts, Joseph Phillips, would say of him:

> Lieutenant Bullis was the only officer ever did stay the longest with us. That fella suffer jest like we all did out in de woods. He was a good man. He was a Injun fighter. He was tuff. He didn't care how big a bunch dey was, he went into 'em everytime, but he look after his men. His men was on equality, too. He didn't stan' back and say, 'Go yonder;' he would say, 'Come on, boys, let's go get 'em.'

Invasion in a Forbidden Land

In the spring of 1873, the time had finally come for the cavalry to strike back. After all, the scouts had the uncanny ability to follow a trail that was weeks old. They could quietly surround an enemy camp before anyone ever became aware that death lay only a few feet away. The military looked coldly at last to the borders of Mexico, then beyond. It had one chance to destroy the base that had plagued the frontier for so long. The military handed the assignment to Ranald Mackenzie, once described by U.S. Grant as "the most promising young officer in the Army." Mackenzie, in 1862, had graduated at the top of his class in West Point. During the War Between the States, he was wounded at the Second Battle of Manassas in the midst of action that would earn him his first brevet for "Gallant and meritorious service." June of 1864, Mackenzie became Colonel of the Second Regiment Connecticut Volunteers in the Sixth Army Corps. When the Union force sought to drive Robert E. Lee from Petersburg, Virginia, he lost two fingers of his right hand to a gunshot wound. The Indians would forever call him "Bad Hand." By the time Mackenzie joined General Sheridan in the Shenandoah campaign, his adjutant said of him: "He had so far developed as to be a greater terror to both officers and men than Early's grape and cannister." As the fighting roared down the valley, Mackenzie was wounded twice and earned his promotion to Brevet Major General. He stood with Grant at Appomattox, taking charge of the surrendered Confederate property. During those three years of constant warfare, Mackenzie had collected seven brevets, been wounded six times, and had risen to a higher rank than any other man in his West Point class.

For a time, after the conflict ended. Mackenzie, an engineer, was placed in charge of rebuilding dilapidated army posts. He wasn't happy. He was never happy unless he was standing in the face of war. Finally, on December 15, 1870, Ranald S. Mackenzie was handed command of the Fourth Cavalry. When General William T. Sherman needed someone to clear the way for the advancement of civilization west, he was ready. He would not fail.

In the summer and autumn of 1871, Mackenzie had led 600 men against the Comanches who ruled over the Staked Plains of Texas. A year later, he rode away from Fort Concho and chased a renegade band to Alamogordo, New Mexico, tenaciously trailing them, refusing to give up, refusing to go back. The Comanches had never faced such an obstinate foe before. In September of 1872, Mackenzie surprised a village on McClellan's Creek in the Texas Panhandle, capturing 130 prisoners and leaving behind twenty warriors slain upon the ground they had chosen to defend. He had lost only one man.

But out on the southern plains of Texas, the atrocities of the Comanche, Lipan, and Kickapoo had become intolerable. By the spring of 1873, their raids had inflicted about fifty million dollars worth of damage on the countryside. Men had been brutally murdered, their women and children kidnapped, their livestock driven into the arroyos of Mexico.

General Sherman wrote General C. C. Augur from Washington, D.C.:

> The President wishes you to give great attention to affairs on the Rio Grande Frontier, especially to prevent the raids of Indians and Mexicans upon the people and property of Southern and Western Texas.
>
> To this end he wishes the 4th Cavalry to be moved to that Frontier, and it will be replaced by the 7th Cavalry to be drawn from the Department of the south . . . The 4th, as soon as it is safe to move, should march to the Rio Grande, and the ninth can be broken up into detachments to cover the Western Frontier and road toward New Mexico.
>
> In naming the 4th for the Rio Grande the President is doubtless influenced by the fact that Col. Mackenzie is young and enterprising, and that he will impart to his Regiment his own active character.

So into Brackettville rode Colonel Ranald Mackenzie and his Fourth Cavalry. Captain R. G. Carter, his aide, looked around at the little town and later wrote:

> Its composition varied somewhat, but there were the inevitable adobe houses, Mexican ranches of 'shacks,' huts, 'jacals' and picket stores . . . Mexican 'greasers,' half-breeds of every hue and complexion, full-blooded descendants of the African persuasion, low down whites and discharged soldiers, with no visible occupation, composed the population, and at night a fusillade of shots warned us that it was unsafe venturing over after dark on the one crooked, unlighted and wretched street — Le Boulevard de Brackettville.

Fort Clark itself held little promise of comfort for the Fourth Cavalry. The troops, according to an Inspector-General's report, weren't particularly well drilled, and their clothing didn't fit properly. The quarters, he said, were "wretched and therefore nothing beyond shelter . . . All except two companies of cavalry are in huts." The guard house was cramped. He couldn't find any place to hold either a divine service or a court-martial. And he worried about the

Invasion in a Forbidden Land

laundresses who lived in "miserable shanties." The soft bread was of "indifferent quality." Fresh beef cost almost a nickel a pound, and the contract price of corn was $1.44, hay $11.95, and fuel $2 a cord. The shops were "indifferent," but the cavalry stables were in good shape. The kitchen in the hospital building was "unworthy of its name." The Inspector-General pointed out,

> The approaches to the Post are not prepossessing. It is surrounded by little mounds of debris and camp-refuse, apparently the accumulation of years. Its condition has attracted the attention of the present commander & he has done something toward its correction. But there remains much to be done before the police can be called even tolerable. The men's sinks with one exception are not in proper order and one (belonging to a Ninth Cavalry Company) was found in a shocking and shameful condition. This is a proper occasion to remark that many officers seem to content themselves with issuing orders . . . and omit the important duty of seeing personally that their orders are executed.

Colonel Mackenzie wasn't really concerned with the appearance of Fort Clark. He did know how to give orders. More importantly, he knew how to follow them. And his were to rid Texas of the Indian threat that came from Mexican soil. Captain Carter even mentioned that "there had seemed to be more or less indifference or indecision by General Merritt, then commanding the Ninth Cavalry, in dealing with these murderous cut throats, bandits, and thievers." Perhaps he was at a loss. Perhaps he had simply given up. The years and the frustrations had definitely taken their toll.

On April 11, General Belknap, the Secretary of War, and General Phillip Sheridan arrived at Fort Clark, ostensibly to conduct a mounted field inspection of the troops. However, that night, they met with Mackenzie at a secret conference in the commandant's quarters.

Sheridan stared at Mackenzie for a moment, then told him,

> You have been ordered down here to relieve General Merritt and the Ninth Cavalry because I want something done to stop these conditions of banditry and killing by these people across the river. I want you to control and hold down the situation, and to do it in your own way.

Mackenzie listened closely, not quite sure exactly what the general had in mind.

Sheridan continued,

> I want you to be bold, enterprising, and at all times full of energy;

when you begin, let it be a campaign of annihilation, obliteration and complete destruction, as you have always in your dealings done to all the Indians you have dealt with. I think you understand what I want done, and the way you should employ your force.

Mackenzie frowned.

The only way he could ultimately destroy the enemy would be to attack his home base, and that lay far in the innards of Mexico, protected by the red tape of international law. Yet, in his own veiled way, was Sheridan telling him to invade a foreign country anyway?

"General Sheridan, under whose orders and upon what authority am I to act?" Mackenzie asked. "Have you any plans to suggest, or will you issue me the necessary orders for my action?"

Sheridan's eyes burned with anger.

He slammed his fist on the table and gestured wildly. "Damn the orders!" he barked. "Damn the authority. You are to go ahead on your own plan of action, and your authority and backing shall be General Grant and myself. With us behind you in whatever you do to clean up this situation, you can rest assured of the fullest support."

He paused a moment.

Then he concluded with a solemn voice, "You must assume the risk. We will assume the final responsibility should any [trouble] result."

Colonel Ranald Mackenzie had his orders.

He only hoped that he understood them.

Mackenzie's days were nervous and uneasy. He knew that he must act and act quickly. Neither Sheridan nor Grant would allow or forgive any hesitation on his part. The colonel selected two ranchers, Green Van and McLain, to accompany the post guide, Ike Cox, on a reconnaissance mission that carried them seventy miles into the interior of Mexico. He wanted them to locate accessible roads, mark any trail that might permit his men to move at night, and try to determine the number of Indians who had hidden themselves away in the Mexican wilderness. He had faith in his spies. They were truthful and trustworthy, and, besides, the two ranchers had long ago grown tired of losing their cattle to the rustling renegades.

Horses were carefully shod, and out in the grazing camps, the men of the Fourth Cavalry were thoroughly drilled. Carbines barked at targets, and the troops spent hours mounting and dismounting and preparing for war.

No one but Mackenzie knew why.

Sometime between two and three o'clock in the morning, on May 17, 1873, Major Clarence Mauck rode into the camp on Piedras Pintos Creek and awoke Colonel Beaumont. "I have orders from Colonel Mackenzie," he said. "Pack up and saddle up." The major had climbed out of his sick bed to deliver the ominous message, and Beaumont was stunned.

He turned to Captain Carter and asked, "What is the meaning of this — where are we going?"

Carter shrugged. "*Quien sabe?*" he answered. "Perhaps across the Rio Grande."

Beaumont shook his head and pondered: *Mackenzie might fight. He was always ready for a good fight.* But not even Ranald Mackenzie would dare violate the laws of the land. Minutes later, campfires lit up the darkness, glowing like eerie fireflies along the banks of Piedras Pintos. By 8:30 A.M., Beaumont's unit had packed up and saddled up for their rendezvous with Colonel Mackenzie on Las Moras Creek.

The colonel rode at the head of two troops from Fort Clark, flanked by First Lieutenant John L. Bullis and his detachment of Seminole Negro scouts. They waited for Troop M from Fort Duncan to join them. The morning dragged on. Troop M had lost its way and did not arrive at all until the one o'clock sun had risen high above them.

Ahead lay the Rio Grande.

And the 400 men, still unaware of their mission, rode away, with wet sponges fitted into their hats to guard against sunstroke or heat prostration upon the prairie. As far as the Mexican spies in Brackettville were concerned, the cavalry had merely embarked on another long and fruitless foray across a dry and thirsty land. They turned back to their hot, flat beer. They hadn't underestimated Mackenzie. But they hadn't read his mind either. They didn't know the man. And that was their worst mistake.

That night, as the horses paused beside the river, Colonel Mackenzie called his men around him. They were indeed going into Mexico, he said. The raid might be unlawful, but it was necessary. Every soldier, every scout, faced risks greater than any of them had ever known before. If they fell wounded, they might be captured and hanged or backed up against an adobe wall to stare into the lethal eyes of a firing squad.

He didn't, however, mention that he was acting without orders or any written governmental authority at all. In his heart, Mackenzie genuinely believed that he had received implied permission from General Sheridan to invade Mexico. He had asked no specific questions, and Sheridan had given him no specific answers. The decision had ultimately been Mackenzie's to make, and he had not shied away from it.

The colonel thought seriously that he would probably be arrested when he returned. Perhaps there would even be a court-martial. He didn't know. Perhaps he had misinterpreted General Sheridan's words. But he couldn't let that bother him. All he knew for sure was that he had been told implicitly to wage a campaign of annihilation, obliteration and complete destruction. Those had been his orders and, so help him God, he would carry them out. Maybe he would become a military scapegoat. Not once did he believe that Sheridan and Grant would desert him.

None of Mackenzie's men refused to cross the Rio Grande. They were confident and full of hope. They would follow him anywhere. The Indians would elude them, escape them no longer. The thought of death did not particularly frighten them. They had seen it before, dancing all around them. And they would witness death again. But men couldn't survive if they lived in constant fear it was waiting solely for them.

As Captain Carter recalled:

> The river was reached shortly after eight P.M., sufficiently dark to cross without being seen . . . Our reflections were only disturbed by the murmuring of the water and the impatient splashing of our animals. All talking was ordered to cease . . . We stemmed the swift current and . . . scrambled over the low but steep bank and into the dense canebreak . . . It was now too dark to distinguish anything but the dim forms of the moving horses and men. We were indeed upon the soil of Mexico, and without further delay, the start was made for a night's ride upon the distant Indian villages.

The Cavalry blindly followed old mule trails that led past the dried chaparral and out into a desolate land. It was a dreary and trackless desert, bounded by foreboding mountains and listed upon aging maps as Terreno Desconocido —the unknown land.

Captain Carter would write,

> . . . away sped the somber troopers, startling the dwellers in the lonely ranches when the dull thunder of tramping hooves rose and fell as the rapid human torrent poured across plains or plunged into ravines. Lights disappeared from dwellings as if by magic, and perhaps many a devoted mother clasped her babe to her breast in mortal terror at this unusual and ominous roar at the dread hour of midnight.

The men rode at a rapid gait.

They must reach the village by daybreak or all would be lost. Surprise was their greatest hope. The horses galloped on through a choking cloud of dust that rose up to blind them as they sought to tame and perhaps conquer the Terreno Desconocido. The mules had been heavily packed, and they sagged to the rear. The column slowed, realizing that only time could defeat them, and the mules were costing them dearly.

In desperation, Mackenzie turned to Carter and ordered him to "tell the troop commanders I'll halt, and give just five minutes to cut the packs loose. Tell the men to fill their pockets with hard bread."

The Cavalry galloped onward.

Their rations were left behind them, lying to rot upon the desert floor.

Fatigue crept upon the men, and thirst gnawed at their throats. Dawn tainted the eastern sky gray, and Carter thought it was "the personification of death — cold, dreary, and hopeless." To him, the faces of the troops themselves were pallid and "corpse-like." The men were weary and their horses exhausted, but still they rode on at a killing gait. The Seminole Negroes stared ahead with ebony faces, and long black hair powdered with the alkali dust of the desert. The soldiers were haggard, suffering from strain and loss of sleep. The cooling winds of the Santa Rosa Mountains gently rolled down those rocky slopes to greet them.

Ranald Mackenzie was worried.

Perhaps he had misjudged the distance from the river to the villages. Perhaps his guides had lost their way. He glanced upward as the sun began to climb beyond the foothills that still lay in the mist of the early morning.

Soon it would be hot.

Soon he and his men would be too late.

The troops paused beside the water holes of the Rey Molina, and washed the dust from their faces and their throats. The Indians were close now. The journey was at an end. Silence was the greatest ally of the Fourth Cavalry, but the noise of the hooves and the jingling of spurs and the rattle of the equipment, said Carter, became almost painful.

Surely someone would hear.

Surely the Indians would be awaiting them.

Across the dry bed of a stream, the Kickapoo village appeared amongst the mesquite and Spanish dagger and chaparral, its lodges quiet and still sleeping.

Captain Carter remembered the scene well:

> A shot, followed by another and a third, then a front line volley, and the grey horses of I Troop in the lead could be seen stretching down the slope upon the villages ... And then there burst forth such a cheering and yelling from our gallant little column, as that Kickapoo village never heard before. It was caught up from troop to troop and struck dismay to the Indians' heart that they were seen flying in every direction ...

Mackenzie's charge had carried with it the surprise that he had hoped for. Within moments, Troop I had stormed into the grass lodges. As Carter recalled:

> Over mesquite bushes, rocks, prickly-pear, and the long, dagger-like points of the Spanish bayonet, dashed the mad, impetuous column of troopers. Here could be seen a horse gone nearly crazy and unmanageable with fright, and running off with its rider, who was almost or wholly powerless to control him. Small, mesquite trees had to be avoided, and what with controlling the men, dodging obstacles over rough ground, and handling our horses, a more reckless, dare-devil ride we never had.

Some leaped from their mounts to fight on foot.

Others chased the escaping Indians across fields of corn and grain and pumpkins. A few began rounding up stock that had been stolen long ago from Texas ranchers.

Mackenzie yelled, "Fire the villages!"

The grass lodges, dry as tinder, burned like powder, crackling amidst the sporadic gunfire. The destruction, if not the annihilation, had been complete.

Renty Grayson, one of the Seminole Negro scouts, caught old Costilietos, chief of the Lipans, with a lariat as he frantically raced through the brush. Sergeant O'Brien, tough and Irish, chased one warrior who suddenly turned and hurled a brass-bound tomahawk at his head. It missed, and the sergeant smiled. "I have you now, you old Spalpane!" he shouted. Calmly, at fifteen paces, he shot and killed the brave. Ike Cox, the post guide, was kneeling beside the stolen horses and grinning. Upon several he had found the brand "IKE." For him, the long ride had been worthwhile after all.

The villages lay in ruin and desolation.

The cruel tragedies of war slowly spilled out before Mackenzie's men. The day began calm then turned deadly. The labors of war tend to lose their glory when the victors, at last, are forced to stare into anguish and the agony that smoulder gray like dying ashes in the eyes of the vanquished.

Invasion in a Forbidden Land

Captain Carter reported:

> I approached the bushes and, parting them, witnessed one of those singular and pitiable spectacles incident to Indian warfare. A small but faithful cur dog was at the entrance of what appeared to be a small cave far under the bank of the stream, savagely menacing our advance. Near him, almost underneath, lay stretched the dead body of a gigantic Indian, and behind him seemed to be more bodies. It was necessary to kill the dog before we could proceed further. The men reaching in, then drew forth two small children, respectively two and four years of age, badly shot through their bodies. One was dead, the other nearly so. Opening the bush still further for more revelations, way in the rear we saw the form of a young squaw, apparently unhurt, but badly frightened. Her black, glittering eyes were fastened upon the group of blue-coated soldiers with a fascinating stare, not unlike that of a snake, expressing half fear, half hatred and defiance. We made signs for her to come out, but, as she refused, she was quietly, and without harm, dragged forth. We thought this was all, but almost covered up under the immense flags, we found still a third child, a girl of about twelve, badly wounded. It was one of those cruel, unforeseen and unavoidable accidents of grim-visaged war.

Mackenzie would report that nineteen Indians had died on that morning of battle. His men argued that more had fallen. Mackenzie never bothered to cross the ravaged field personally or to count the bodies that sprawled among the ruins. He simply rounded up his prisoners, numbering between forty and fifty, herded together 200 stolen horses, and turned again toward Texas.

An angry sun scaled the prairie around them. The faces of the Mexicans they encountered along the trail were malignant with scowls of hatred. None of the Fourth Cavalry felt safe, and their eyes searched the horizon from which death would come. Perhaps their long march had been discovered, perhaps the Mescalero Apaches from Zaragoza — longtime allies of the Kickapoos and Lipans — would be waiting in ambush. Sleep taunted them. The day slowly faded away beyond the sharp ridges of the Santa Rosa Mountains. The soldiers became irritable, then morose as strength ebbed out of their weary bones. Nerves were worn almost threadbare. For many, the rising of a yellow moon signalled the beginning of their third straight night without rest.

The officers urged their men onward with gentle persuasion and violent threats, pushing them and shaking them as they slumped wearily in their saddles. Sleep would be relief. Sleep could mean the death of them all. Only the Seminole Negro scouts kept alert throughout the night of strain and bewildered minds. On more than one occasion they saw shapes of the enemy hiding in the shadows, trailing the soldiers but not attacking, unaware how tired and vulnerable the troops really were. The men bordered on collapse, but still Mackenzie pushed them through the chaparral and on toward the Rio Grande. He was intruding upon a land where the law had forbidden him to go. Mackenzie didn't want to tarry. He had tied his prisoners to their horses so they wouldn't weaken and fall, delaying his journey. He led his Cavalry on down a gentle ridge at daylight, through a dense thicket, and the men washed themselves again in the muddy waters of the Rio Grande. As Captain Carter remembered, "The long, interminable night of horror, of night-mare, had passed...All faces wore that dull gray, ashy, death-like appearance, indicative of overworked nature and the approach of exhaustion and physical collapse." One of the wounded Indians died beside the river. The children, half-naked, streaked with dust and sweat, slept upon their horses, their faces stained with warpaint and tears. Then, when on American soil again, Green Van brought several buckets of throat-burning Mexican mescal down to the weary troops. He thought it might refresh them. Mackenzie ordered the home-distilled fire spilled out on the ground. The Indians hadn't killed his men. But the mescal, when poured into empty stomachs, might.

A few miles away from the Rio Grande, the men at last stopped and bivouacked on open ground, surrounded by a wall of dense chaparral. As the officers sat together, Colonel Eugene Beaumont suddenly asked Mackenzie if he had been given any special orders to lead his command into a foreign country.

Mackenzie shook his head.

Beaumont was stunned. "Then it was illegal to expose not only the lives of your officers and men, in action," he said, "but, in event of their being wounded and compelled upon our withdrawal, through force of circumstances, to be left over there, probably to be hung or shot by a merciless horde of savage Indians and Mexicans."

"I considered all that," Mackenzie answered.

"Your officers and men would have been justified in refusing to obey your orders, which you now admit to being illegal, and exposing themselves to such peril," Beaumont told him.

Anger began to redden the face of Captain N. B. McLaughlin. "Beaumont is right," he snapped. "And had I known that you had no orders to take us over the river, I would not have gone."

Mackenzie's voice, Carter recalled, was firm, crisp, and deci-

Invasion in a Forbidden Land

sive. "Any officer or man who had refused to follow me across the river I would have shot," he barked.

"That would depend, sir," replied McLaughlin, "on who shot first."

By noon on May 21st, Mackenzie and his raiders had returned again to Fort Clark. No one knew where he had gone. No one knew why.

Major J. K. Mizner, relying on a report he had received the day before, had already sent the following letter to the Assistant Adjutant General of the Department of Texas in San Antonio:

> A dispatch received at this Post last evening, brings information of a brilliant success in an attack made upon two villages of Kickapoo and Lepan Indians by General Mackenzie and the 4th Cavalry under his command.
>
> General Mackenzie with Cos A, B, C, E, I and M 4th Cavalry and 25 Seminole Scouts struck a camp of the Kickapoo and Lepan Indians about 80 miles from this Post, early on the 18th inst, having marched all the night previous. Killed 19 Indians Wounded 2 and captured one Buck a former Chief of the Lepans took 41 women and children prisoners and has already sent in over fifty captured ponies. Besides destroying two villages with all their accumulated property.
>
> All this was accomplished between noon on the 17th and daylight on the 19th and with a loss to General Mackenzie's Command of but three men wounded, one it is supposed mortally.
>
> The march (over one hundred miles) the fight, destruction of the villages and the return to Camp with all the captured stock and prisoners was accomplished in a little over forty hours.

Mackenzie's reputation had been at stake.

Success, he believed, exonerated his decision to invade Mexico. In the general's own report, he said of his men:

> All of these officers acted handsomely and deserve consideration, and every soldier showed after their terribly hard ride a creditable eagerness to attack.
>
> It was good fortune however of Captain McLaughlin with his Company 'A' 4th Cavalry to be in the advance of the column and I feel called on to mention the very gallant manner in which himself and Lieut Hudson led the Company therein the men of which acted gallantly to the extent of rashness.
>
> I also wish to mention Lieut Bullis with the Seminole Scouts who charged under the command of that gallant officer very well.
>
> I wish it understood in making special mention of these officers that others very probably acted quite as handsomely, but

from leading the advance, these attracted notice more especially. In an Indian fight Officers and men soon get so scattered in the pursuit of Indians that it is perfectly impossible to give to each his proper credit . . .

The command being most of the time at a trot or a Gallop and consequently for two days the men were entirely without rations except a little hard bread in their pockets, yet there was no complaining . . .

No Horses gave out on the Scout, and the only Horses lost were two shot in action and one or two which died from exhaustion in the chase . . . I have omitted to state that three Villages averaging from fifty to sixty Lodges were destroyed. They appeared to be well supplied with stores, including ammunition.

Private William Pair of Company I lost his right arm near the shoulder. Private Leonard Kemppenberger of Company E was slightly wounded in the face, but the injury was not bad enough to keep him off duty once he was back at Fort Clark. Private Peter Corrigan of Company D suffered wounds that took his life.

The land, once considered harsh, was gentle compared to the squalor of mankind who fought to hang on to their bramble brush homesteads. Hans Mickle learned in a hurry. He spent a long, chilling night in Eagle Pass, trying to make his way across Texas. In 1883 he wrote:

> I inquired at the depot for a lodging house, and was directed to a tent with a light in it. Entering, I found it a saloon, occupied by a man on a cot on one side of the stove, and a man lying on some benches behind it . . . The man I asked for lodging said that his sleeping tent in the rear was full and I must do the best I could which was sit by the stove. After a while I laid down on a gambling table in a further corner, but getting cold had to hug the stove again and so I passed the night taking short naps on the gambling table and nodding in the chair . . .
>
> At 8 o'clock the Brackett stage came around and it was boarded for an eleven mile ride. I have traveled in almost every way and in all sorts of company, but the personnel of this stage load could not easily be duplicated for its 'mixed' character. The driver was an Irish discharged soldier, rather fond of the 'crathur.' Then we had a drummer, two gamblers, a Mexican p[r]ostitute, the deputy sheriff, with the Mexican wife murderer, who had shot his wife in Del Rio the day before and Hans Mickle.
>
> To while away the tedium of the ride the gamblers opened a monte game and won several dollars from the drummer and Mexican woman. The day was raw with a stiff norther, and nobody was sorry when a halt was made in Brackett.

Perhaps Brackettville had seen better days. Perhaps, for a time, prosperity had deserted the town.

But Mickle, in spite of the hard times, was duly impressed.

He wrote: "The post at Fort Clark is its main stay-by, and the bi-monthly payments of the troops are hailed with rejoicing. The most serious question that agitates the Brackettite is the removal of the troops and the abondonment of the post."

That was a problem that would long haunt the dusty little town.

Mickle, however, believed, "A better place to quarter troops cannot be found on the frontier, as there are already ample accommodations, costing several hundreds of thousands of dollars, it would appear like a bad move to take away the troops from this place while the Rio Grande is our national frontier."

He promptly rode on to Del Rio, looked around, and boldly predicted, "This undoubtedly is one of the 'future greats' of Western Texas.'"

General Ranald Mackenzie wondered what reprisals he might face. He waited for a reprimand or worse. General Philip Sheridan did not back down on his vow to support the Indian fighter in whatever he might take. In one letter he wrote:

> There is in my opinion only one way left to settle the Mexican Frontier difficulty, that is, to cross the Rio Grande and recover our property, and punish the thieves.

Two weeks later, he wrote,

> There should be no boundary line when we are driven to the necessity of defending our lives and property against murderers and robbers. If the Government will stand by this action of Col. Mackenzie the troubles on the Rio Grande frontier will soon cease.

And a third letter to the Secretary of War pointed out,

> If the government will stand firm we are near the end of all difficulties on the Mexican frontier. For twenty years there has been no safety for life or property on the Rio Grande, and at last when we are driven by a necessity which is the foundation of all law, namely that of protection of our lives and our property, to cross a very crooked river to exterminate the murderers and robbers, we must in all reason be sustained. There can not be any valid boundary line when we pursue Indians who murder our people and carry away our property. If the government will say that there shall be no boundary line . . . there never will be another raid on the Rio Grande.

From deep in the interior of Mexico came word that the Indians had banded together and applied to the alcalde at San Fernandez for permission to launch a full-scale attack on the American border. The alcalde refused, and the chiefs carried their grievances to the governor himself. They passionately urged Mexico to join them, sending troops and weapons, to conquer the Texas border. A scout, Bill Donovan, reported that he had seen a war party traveling east, and each man carried a rifle and was riding on a fine horse.

Mackenzie acted quickly. He wrote William Schuhardt, American Consul, agent in Piedras Negras, gambling that the governor of Coahuila would see the letter, and he warned:

> The Indians can be well assured that they have not got nor can they get enough men to hurt me, for every man of mine I will hurt many more Indians.
>
> You can assure the Alcalde of my friendly feelings toward Mexicans . . . My desire is to go hand in hand with the Mexican authorities in suppressing all frontier lawless acts, and should the Comanche or other tribes depredate in Coahuila, that I would be willing not only to follow and punish them on this side of the river but also on the other; but that now I wish and believe that all good Mexican citizens should give me every assistance in putting an end to the terrible outrages which threaten both sides of the river . . .

The Indians knew that Mackenzie's words were not idle ones. Diplomatically, he was offering to fight for the Mexican government as well as his own, aiming only for the annihilation of the renegades who had left behind far too many graves in the scorched Trans-Pecos soil.

They feared Mackenzie.

They respected him.

The raid never came.

The legislature of Texas promptly passed a resolution to show their support for Mackenzie and the Fourth Cavalry who had trampled on sacred Mexican soil to bring them a measure of peace:

> Whereas reliable information has been received that General Ranald Mackenzie of the U.S. Army with the troops under his command, did . . . inflict summary punishment upon the band of Kickapoo Indians who harbored and fostered by the Mexican authorities have for years past been waging predatory warfare upon the frontier of Texas, murdering our Citizens, carrying their children into captivity and plundering their property:
>
> Therefore, Resolved — by the Senate of the State of Texas. The House concurring. That the grateful thanks of the people of

the State and particularly the Citizens of our Frontier are due to General Mackenzie and the officers and troops under his command for their prompt action and gallant conduct in inflicting well merited punishment upon these scourges of our frontier . . .

For a time, an uneasy peace spread across the land, alway under the guarded eyes of Ranald Mackenzie.

6

The Long March North

The sentries of Fort Clark kept their eyes turned toward the land from whence war would come, if it came, and the soldiers braced themselves for an Indian onslaught. The Kickapoos were demanding that their women and children be returned to them, and the army was making steady preparations to march the prisoners northward to Fort Gibson in the Oklahoma territory. The Mexican government, to no one's surprise, had stormed out in vocal indignation about Ranald Mackenzie's foray into its homeland. They would retaliate, the rumors said. They would supply arms and ammunition to any tribes who dared to ride against the arrogant, law-breaking Cavalry of Fort Clark, or so the gossip spread.

For a time, it appeared that both Colonel Thomas G. Williams and Henry M. Atkinson — special U.S. commissioners to Mexico — would be detained, maybe even locked away in a forgotten prison cell. After all, they had been meeting with Mexican officials to negotiate the removal of Kickapoo and Potawatomi Indians to a United States reservation when Mackenzie began his raid on foreign terrain. The officials promptly blamed the commissioners and threatened to make them pay personally for the Cavalry's sins. Mackenzie's reaction was expected. "If they do imprison Atkinson & Williams," he wrote, "I believe we ought to go unto them at once ... My disposition would be to take every Cavalry man I have got and try to make the Mexicans at Santa Rosa see how wrong it was to worry them." The Mexican government didn't take any chances. It wouldn't be wise at all, they decided, to raise Mackenzie's ire. They sent Atkinson and Williams home.

Mackenzie wanted to take the offensive.

It wasn't his style to sit and wait and wonder what his fate would be. He didn't believe in giving the Indians a chance to regroup and gain strength. He wanted to strike again while the tribes were scattered in chaos and confusion. He wrote General Augur:

> If the Mexicans do not make the Indians now come out the Authorities may just as well make up their minds to have war. The Indians can easily move their Camps so far back into the Mountains and so far away from the river that they can be reached by operating for a considerable time on the other side of the Rio Grande.
>
> I believe that if the Mescaleros or Lepans could be hit pretty hard that the Kickapoos would conclude to come out. Opportunities for striking Indians are not very frequent, and if there is a good chance of attacking these people, I should like to to so as soon as it can be done without interfering with the Commissioners. The government might however just as well determine that either the Mexican Troops must make the Indians come to this side, must allow our troops to go over publicly to go regularly for months after the Indians, or that we will have to fight the Mexicans. These frontier Mexicans have a ridiculous idea because the opposite side of the Rio Grande is here more thickly settled than this side that they are stronger than we are.
>
> It is my belief that if the Indians were got rid of that the Cattle stealing would very soon die out.
>
> I wish to say in closing this letter that personally I am very much opposed to a War with Mexico if it can possibly be avoided, but this plundering might as well be stopped now as later, in fact it is a great deal best to stop it now.

Seven days later, Mackenzie got his answer.

General Augur wrote: "Colonel Mackenzie has been ordered not to cross again into Mexico, without further provocation."

General Sheridan agreed. "I have directed General Augur to tighten up a little on Col[onel] Mackenzie."

General William T. Sherman quickly concurred with them that Mackenzie should use some caution if not restraint. He wrote: "The sending of spies into Mexico, to hunt up camps with a view to a future attack may lead to dangerous complications whereby we may commit a national wrong."

For the moment, the Mexican government had been appeased. Yet, Colonel W. R. Shafter told Mackenzie at Fort Clark that he had learned Indians were receiving powder and caps from authorities south of the border in order to kill deer. Perhaps the bullets would be aimed at soldiers instead. As far as he was concerned, the army should no longer make any distinction at all between Indian and Mexican thieves.

Both were making life miserable upon the prairie.

A band of Indians slipped into Sabinal Vanio and raced away with a dozen of Charles Durlon's best horses. Mackenzie reported to Augur that several groups of Indians had been spotted throughout the countryside and that five different scouting parties were trying to track them down. Lipans who spoke Spanish camped three miles outside of San Antonio, then rode quietly into town to buy arms and ammunition. In a letter to Mackenzie, Colonel Shafter wrote:

> Williams (one of the commissioners) also says the belief is general that the Kickapoos have an understanding with the Comanches and Kowas, and that as soon as the women and children are in a place of safety they will retaliate for Remolina by a big raid made only to slay . . . Williams says the Indians are in a generous rivalry trying to break each other's necks in their haste to get to this side of the Rio Grande.

The winter had rolled its chilling winds across the prairie when Lieutenant Charles L. Hudson led his command toward Kickapoo Springs on the West Fork of the Nueces River. He had left his wagons behind. They were useless upon a terrain so rough that only pack mules could negotiate the severe arroyos and rocky canyon walls. On the east prong of the Nueces, Hudson found thirty-one Indian ponies standing alone with no sentinel to guard them. The renegades had been careless, he thought, and that — in the West — could be the most serious, most deadly mistake of all. The lieutenant hid his men behind the rocks and watched for the Indians' return.

For two days, Hudson waited, until a scout returned with word that the warriors had been seen at the San Antonio crossing of the Nueces. The lieutenant immediately marched his troops away, moving for eight miles, finally locating the renegades on a far ridge, resting their horses. Hudson knew they would not escape him.

He reported to Fort Clark:

> They opened fire with carbines at about 400 yards . . . I then saw some of them running upon which the command mounted and charged. The country is very rough, rocky, and cut up with ravines; the hills so steep that a horse must walk up. It had been rain-

ing and the rocks were very slippery, altogether unfavorable for cavalry. The pursuit was kept up as long as an Indian could be found, then the command returned to the Indian herd . . . Fifty two ponies and mules, a lot of saddles, bridles and blankets were captured in the affair.

Only one soldier had fallen, Private George Brown. He had a flesh wound, one that would definitely not keep him out of action for long.

Nine of the Indians lay dead upon the field. Lieutenant Hudson pointed out, "Others were known to have been wounded but got away, or so hid that they could not be found, the cedarbrakes being very dense."

One of the slain was the son of the Kiowa Chieftain, Lone Wolf.

Down beyond the wasteland of Torreno Desconocido, Lone Wolf wallowed in despair and in mourning. He cut off his hair, killed his horses, and burned his wagon, his buffalo robes, and his lodges.

A detachment of soldiers left Fort Clark and traveled northward, led by Picayune John Wood, the diminutive and aging Seminole Negro scout. He had hopes of overtaking the band of Comanches and Lipans that were riding rapidly ahead of them. the troops had given up. After all, they had spent so many days chasing Indians and so few days ever finding them. More than likely, The renegades would simply fade into the evening shadows and be gone. The soldiers had no hopes of catching them at all. Yet the wrinkled, scrawny little Seminole Negro scout was trying.

The troops laughed.

They ridiculed Picayune John.

The Cavalry was tired. John stayed alert. The Cavalry reasoned that the Indians were running free too many miles ahead of them. John kept finding recent tracks and locating fresh signs left behind by the Comanche and Lipan warriors. Who could believe the old black man? His eyes were bad. His mind wasn't nearly as sharp as it had once been.

The troops laughed.

They ridiculed Picayune John.

That night, as the men made camp, there was no special detail named to guard the men. John wasn't pleased at all. He didn't trust the night or the mysteries that it held. So he chose himself to keep watch far outside the camp boundary. At midnight, John sounded the alarm, and the entire camp came to life, ready for battle. In the distance, the men could see, so they thought, the approaching shapes of an Indian band, and they fired wildly and fearfully into the darkness. There were no Comanches, no Lipans creeping toward them. Their targets had merely been shadows, made by the campfire as its light flickered through the resin weed. Those weaving, bobbing sinister shapes hadn't been humans at all.

The old scout swore that he had seen a warrior crawling in the darkness toward him. He hadn't been mistaken. The enemy did lie in wait.

The troops laughed.

Again they ridiculed Picayune John.

In the dim gray of early daylight, John led some officers to places where Indians had left signs of their midnight watch. No one believed him. No one ever took the old man seriously. All day, he scouted the prairie, returning after dark, convinced that an attack would come. No one listened. His words fell on deaf ears and damp ground. He was angry and useless, and John watched the darkness, and he waited, alarmed by an uncanny instinct that understood both the land and the Indians who rode there.

As the first streaks of dawn splashed across the sky, the aging, battle-scarred scout abruptly shook the shoulder of his sleeping commanding officer and said simply, "They come."

The Indians were upon them.

The camp had been surrounded by the Comanche and Lipan braves, but they met a deadly volley of rifle fire. John's warning had come in time. The campfires were quickly blacked out, and the soldiers drew its defense closer together for strength. By sunrise, the Indians had slipped away, dragging their dead and their wounded with them. A massacre had been at hand.

No one laughed.

No one ridiculed Picayune John anymore.

The Seminole Negro scouts had first proved their worth when they rode with Mackenzie into Mexico. They became invaluable during those constant hit-and-run skirmishes that appeared and vanished as quickly as whirlwinds upon the prairie. The Indians were no longer sure of their escape. They could easily outdistance and usually outsmart the army on land it knew so well. Now the Seminole Negroes hung as close to them as a scent on a skunk. The renegades couldn't lose them and almost never outfought them.

In one battle, Renty Grayson's mount was shot out from under him by an Indian sniper's fire, but he never stopped nor looked back. He ran headlong into the fray and walked away from it without a scratch. In another skirmish, Grayson dived behind a tree as a bullet plowed up the ground around his feet. He lay still, and he waited. Renty Grayson had patience, and it kept him alive. He quickly surveyed his situation, realizing that he was being held at bay by only one Indian, armed with one cap and ball rifle. The scout carefully placed his tattered old campaign hat onto his ramrod and eased it out into plain view. The Indian fired again, and the hat was spun off the stick by the impact of the bullet. The renegade would not have time to reload again, and Renty Grayson knew it. He charged, and he left a dying Indian in the dust behind him as he rode away. From Fort Clark, he journeyed to Kickapoo Springs and discovered a group of white cattle rustlers who had hired a band of Indians to steal cattle for them. The outlaws had learned that troops from the Fort were riding hard their way, intent on driving them from the Trans-Pecos or burying them there. They lay in ambush, and the Cavalry would have had no chance at all. Renty Grayson found the column of soldiers with the same uncanny ability he had used to track down the rustlers. The troops were saved from riding, in ignorance, into a lethal hail of gunfire.

Elijah Daniel served as Chief of the Seminole Negro scouts at Fort Clark. Jim Bruner became known as the army's "best trailer," according to Major Henry C. Merriam. Julian Longoria was dispatched into Mexico to bring back an Indian for questioning. But, as the report read, he "unfortunately broke the Indian's neck and brought back only a dead body." He refused to return to Fort Clark empty-handed. William Miller, it was said, was a Seminole who looked like a white man and acted like an Indian. With Lieutenant John Bullis, he crept one night into an Indian camp and stole thirty horses. The Seminole Negro scouts were definitely making their mark on the Texas frontier.

But, in the winter of 1873, they found that the promises of the U.S. government were just as hollow at Fort Clark as they had been on the sands of Florida. The army needed them. The government ignored them. To the Cavalry, they were necessary. To the government, they were a nuisance.

The army praised them.

The government lied to them.

The Seminole Negro scouts found themselves again without a home and without a country to rightfully call their own. Three years earlier, they had been promised land grants if they would only serve as scouts. Still they squatted with their families on a military reservation. They had grown weary and impatient waiting for the sacred acreage to be placed in their name. Land would give them permanence, a sense of belonging. They wanted it. They deserved it. They had earned it on battlefields in both Texas and Mexico.

What had happened? That was a question that burned deep in

all of the scouts. Had they risked their lives, had they again uprooted their families for nothing?

The War Department's answer was a simple one. We don't own any land that we can give you, it said. Besides, the Indian Office had closed the rolls on the Seminole tribe back in 1866, shutting out all who had chosen to reside in Mexico. The Indian Commissioner bluntly told them, *You should not have left the United States in the first place. And since you did, you should have stayed in Mexico.*

There was no gratitude for them.

There was no land.

In 1870, the government had provided rations for all of the Seminole Negroes who had agreed to return to American soil. Now, only the enlisted scouts received food. All others were cut out of the rations. As a result, the wages of about fifty scouts were spread around to feed and clothe the entire community. Hunger again became a constant companion. The Seminole Negroes had lived poorly, but now barely at all. The women searched for work in Brackettville, and the children cried. Crops, even during a good year, were sparse. Usually they died beneath a cruel sun.

Both Colonel Mackenzie and General Sheridan endorsed the scouts claim for land. Lieutenant Bullis stood behind them, declaring that the Seminole Negroes were "fine trailers and good marksmen." Colonel Edward Hatch of the Ninth Cavalry described them as "brave and daring, superior to the Indians of this region in fighting qualities."

The officers who had ridden into battle with the scouts knew them best.

They understood their needs and their dreams to own a parcel of land, even in a worthless desert.

The words, the support, of the officers were quietly filed away and forgotten.

On December 10, 1873, John Kibbitts and John Horse made their appeal in person to Brigadier General C. C. Augur, commander of the Department of Texas, in San Antonio. His was a compassionate ear. He believed in them and in the justice of their cause. But his hands were legally tied. If a formal treaty had ever been signed with the Seminole Negroes when they were holed up in Mexico, no one could find it. General Augur sadly shook his head. I'm sorry, he said, but there is nothing I can do for you.

On the morning that the scouts' appeal was turned down, on the morning that their pleas had fallen in vain, nine Seminole Negroes were riding bravely into action with the Fourth Cavalry at Kickapoo Springs. In the swift, but decisive, battle, nine Kiowa and

Comanche warriors died, and sixty-one horses were captured. When the dust had cleared and the smoke had faded from the prairie, the scouts — their heads held high — came riding back to land that was owned by someone else, that would never belong to them.

There would be cries of anguish.

There would be the ache of disappointment.

A few Seminole families, weary of the promises and the lies, packed up and walked away from Forts Clark and Duncan, heading back to the isolation and the safety of Mexico. Some scouts, in disgust, deserted the army. Even Picayune John left. But as the Seminole Negroes would say, "John Wood didn't really desert. He was an old man and didn't know what the army was all about. He just got tired and went home."

Most of the scouts stayed on.

They were stubborn. They were proud. Someday, they said, the wrongs would be righted, justice would be done, life would be brighter again, and the land would be theirs.

They masked their bitterness behind stern ebony faces and rode again in search of Indians whose raids had branded the Texas frontier with tribulation. For a time, the attacks had decreased steadily; a new, though uneasy, peace fell along the Rio Grande. For days, then weeks, the horse soldiers would find no tracks of the Comanche or Kickapoo or Lipan renegades at all. Ranchers seldom reported the loss of any livestock. An unfamiliar calm pervaded the settlements that had been scattered across the brush country.

In General Augur's annual report for 1873, he stated his own opinions and his concerns about the Indian threat in Texas. Writing of Mackenzie's raid, he pointed out:

> The excitement on the frontier consequent upon this affair becoming known was very great for a few days, but when it was clearly understood that it was simply a deserved punishment inflicted upon persistent murderers and thieves the better class of Mexicans were satisfied. They reasoned that so long as the Mexican government failed to prevent these atrocities and punish their authors, it was but right our troops should do it . . . Our Government having sustained Col. Mackenzie's action the Indians felt there was no longer absolute safety for them in Mexico. They must either return to the U.S. or remain in Mexico and cease their depredations on the Texas frontier, or else remove their homes into the mountains beyond the reach of our troops . . .
>
> In addition to sending the 4th Cavy. to Fort Clark, I increased by four companies of 9th Ca. the Cavalry on the lower Rio Grande, and the cattle stealing in that district has nearly ceased

this year. The troops there have also been useful in assisting the Civil authorities in breaking up some resident bands of robbers.

As far as General Augur was concerned, the serious threat of Indian attacks in the Trans-Pecos had passed. There would, of course, be occasional raids by those who only knew how to survive by stealing. Some might die upon the prairie. Some men would risk their lives to take property that didn't belong to them. Some would lose theirs trying to defend it. That had always been the curse of mankind. The general was far more worried about the news he had been receiving from the Texas Panhandle. He wrote:

> On the Northern frontier of Texas, the raids by Indians and outlaws have been numerous — much more so than the previous year, and I regret that thus far they remain unpunished. These depredations are committed by Indians belonging to the Ft. Sill reservation who readily find refuge there from pursuit and punishment. It is by chance alone that opportunity occurs for troops to meet or overtake them. However vigilant the troops may be, the Indian on his raid is more so — however well mounted the trooper, the Indian has three mounts to his one. Whatever care may be taken to secure the best arms and ammunition to the troops, the Indian finds means through his friends to be fully his equal in that respect, and in his entire base familiarity with the country vastly superior.
>
> There are objections of course to permitting troops to follow Indians to their reservations as now managed, from the liability of innocent Indians being punished for acts of guilty ones; but when Indians guilty of atrocious murders of women and children are known, why are they not arrested and punished like other criminals? Why are they permitted to swagger and boast of their exploits in and about the agencies, demoralizing other Indians, and creating and spreading among them a contemptuous spirit for the authority of the government . . .

The Red River country, by the summer of 1874, was caught in the tenacious wrath of hostile Comanche, Kiowa, and Cheyenne outlaws. Lone Wolf was wrought with anger, vowing to avenge the death of his son at the hands of Lieutenant Hudson. Both Satanta and Big Tree, Kiowa chieftains, had been released and were urging their people to fight. Those Comanche women and children, captured by Mackenzie during his raid into Mexico, had been banished north of the Red River and were planting seeds of hatred in those who milled about reservations. The buffalo, the tribes knew, were quickly being exterminated from the Southern prairies, and few trusted the government to care for their needs. They had depended

The Long March North

on the government before and gone naked and hungry. The buffalo was their hope for survival, and it was quickly vanishing before them. A traditional way of life, handed down by their fathers, was passing as the night, and the Indians were reaching out in vain for the last remnants of their tattered heritage. Ishatai, a Comanche warrior-medicine man, and Quanah Parker led an attack of Comanche and Cheyenne braves on a small party of buffalo hunters huddled together at Adobe Walls. The war, long in brewing, had begun.

Colonel Ranald Mackenzie had ridden from Fort Clark to Fort Griffin when his orders arrived. General Augur told him:

> ... the object of the proposed Campaign against the hostile Cheyennes, Comanches, Kiowas, and others from the Fort Sill reservation, is to punish them for recent depredations along the Kansas and Texas frontiers, and you are expected to take such measures against them as will, in your judgment, the soonest accomplish the purpose ... In carrying out your plans, you need pay no regard to Department or Reservation lines. You are at liberty to follow the Indians wherever they go ... you will impress upon your subordinates when acting away from you, that in a hostile Indian Country, there is never a moment when it is safe to relax in vigilance and precautions against surprise. A Commander against hostile Indians is never in such imminent danger as when fully satisfied that no Indians can possibly be near him.

Colonel Mackenzie and the Fourth Cavalry turned northward toward the Staked Plains, led by thirteen Seminole Negro scouts from Fort Clark. The trail would be a crooked one, leading from one watering hole to another. As the gray of September draped around the troops, snowflakes and a chilling rain swept into the Panhandle, chasing the men into the shelter of small ravines and brakes. Horses gave out and had to be shot. The cavalry didn't dare leave them behind for the Indians who took every mount they could find. The rain became a downpour, a wet norther whipped out of a bruised sky, and the troops pushed on through a quagmire of mud to the rim of Tule Canyon. Captain R. G. Carter recalled,

> It was a weird sight — this long, dark column of mounted men moving almost silently over the thick, short buffalo grass, which deadened all sounds ... Strong guards were posted about the horses ... 'sleeping parties' were placed among the horse herds now lariated out, to guard against any surprise or attempt to stampede our animals. Everybody slept with their boots on, ready on the instant of any alarm for immediate action.

On the night of September 26, the attack came. At first, a hard-riding band of Indians came over a small rise and charged wildly into the camp, firing and yelling and hoping to chase away the horses and leave the Cavalry on foot. They failed. Time and again, the warriors probed the outer reaches of the soldier's defense position, finally crawling into a long, winding ravine to snipe away at any trooper who might move. The sporadic gunfire lit up Tule Canyon throughout the night, and just before daylight, the men quickly mounted and galloped into the midst of the renegades. The warriors fled, leaving fifteen dead on the rim of the canyon, including Woman Heart, a noted Kiowa Chief. Mackenzie's Tonkawa scouts ran down one of the Indians and took his scalp, leaving the dying brave as a grim warning to any who might test the horse soldiers again. The Indian loss, perhaps, would have been greater. But as Mackenzie reported: "The Indians saw us coming down a very long and rough hill and had time to get away but lost almost all their stock."

Within a week Satanta and thirty-seven warriors surrendered to Colonel Thomas Neill of the Sixth Cavalry, turning over thirteen rifles, three pistols, eighteen bows, and four lances. Satanta said simple, "I am tired of fighting and do not want to fight anymore. I came here to give myself up and do as the white chief wishes." But Lone Wolf swore he would never give up or be defeated until his son had been avenged.

And Mackenzie with the Fourth Cavalry swept on across the Llano Estacado. His expedition was described by a special correspondent in the October 16, 1874 edition of the *New York Herald*.

Of the Staked Plains, it was written:

> Not many years ago it was designated as the Great American Desert upon the maps, and supposed to be a vast sandy wasteland like the Sahara of Africa. Now, the geographers inform us, it is 'an elevated table land without wood or water, across which a wagon route was formerly marked by stakes.' The plain rises abruptly four or five hundred feet above the surrounding country . . . covered with a heavy growth of buffalo grass . . . and which, as may will be imagined, is the winter home of countless herds of bison, who come down from above the Canadian, and of well mounted bands of Comanches and Kiowas, who, after drawing their annuities and supplies upon the reservation, make this a base form which to go on their raids after Texas herds and scalps. The eastern and southern sides of this great table land are gashed and seamed by a succession of canyons and arroyos, the raggedness and grandeur of which are beyond description. The edge of the Staked Plain is chaos itself.

The Long March North

Into the chaos rode the Fourth Cavalry, the men from Forts Richardson, Duncan, and Clark. To the Seminole Negro scouts, it was a strange land, unlike any they had ever seen before. It didn't slow them down. They could follow Indian tracks anywhere, across any kind of terrain, and the great canyons didn't confuse them at all.

The unnamed correspondent wrote:

> The command moved north, along the edge of the Staked Plain, on the morning of the 19th. On the 20th a party of four spies, who were one day's march in advance, were discovered and attacked by about twenty-five Comanches, and, after a short, sharp fight, were obliged to run for their lives. One of them had his horse shot from under him, but he succeeded in getting safely upon another old animal, and the whole four reached the Seminole Tonkawa camp in safety. Their pursuers took the alarm, and run off their herd of about one hundred animals in a southerly direction with all haste. Gen. McLaughlin was at once detached with part of the cavalry in pursuit, and he followed the trail more than thirty-five miles before dark, obliging the Indians to drop several horses and abandon considerable camp equipage in their hasty flight . . .
>
> On the 26th a few Indians appeared and skirmished with the outposts at Tule Canyon . . . At 10:15 P.M. several parties of from ten to thirty Indians dashed up on different sides of the camp, firing in among the men and horses, and yelling like the blood-thirsty demons, as they are —peace commissioners and Indian agents notwithstanding— in a vain attempt to stampede the stock . . .
>
> Before it was fairly light the Indians, who evidently had no idea of the strength of the column, came boldly out on the hill for a fight, to the number of about 150. Captain Boehm, with his company, and Lieutenant Thompson, commanding the scouts and Indians (who is known to his messmates as 'Hurricane Bill,' from the plan of his attacks) were ordered to charge and drive them from the vicinity, while the rest of the companies were saddling and packing . . .
>
> In the most gallant manner, losing no men, but killing one Indian, Hurricane Bill and his scouts acted as flankers on each side, killing another Comanche without loss to themselves. This Indian, one of the finest looking fellows your correspondent ever saw, was going up the side of an arroyo when a Seminole jumped down from his saddle, and taking deliberate aim, killed the Comanche's horse, who, being thus suddenly dismounted, started off at a great rate across the prairie on foot. One of the Tonkawas then ran his horse upon him and gave the *coup de grace* with his sixshooter, the pistol being so near his head that the powder burned his skin.
>
> The Comanches scattered in every direction, making no more fight . . . But the cavalry did not follow the trail, of course. That device is getting a little threadbare; every step taken on that trail,

which was made with intent to deceive, would have carried the command away from their villages, and villages were what the column was after; for if their herds and homes are struck it tells very seriously, but if you stike a moving party of warriors, a possible dead Indian or two represents the sum total of the harm done then . . .

As soon as it was dusk the command saddled up and started off almost due north, at a great pace, and traveling all night, arrived on the brink of the big canyon near the head of the Red River just as day was breaking. This canyon is from 500 to 800 feet deep, and at this point about half a mile wide. It was here that Gen. Mackenzie expected to find their villages, and he was not disappointed. Looking far down into the valley beneath their feet the troopers could see the lodges stringing for miles down the river, and the Indian herds grazing in all directions, appearing to one looking from such a height like chickens and turkeys. Officers and men saw their enemies before — or rather below — them, but how to get to them was the question. At last a narrow, dizzy winding trail was found, such as a goat could hardly travel, and the cavalry started down the sides of the precipice. The Comanches took the alarm at once, and began their retreat.

The descent took nearly an hour, but at last the soldiers were down, and Thompson, with his scouts, and the companies of Col. Beaumont, with Captain Boehm, charged the village with such impetuosity that the Indians ran in every direction. After a short but sharp fight, the companies charged on down the canyon through the village, which proved to be scattered along for about three miles, to its end before halting. It was a running fight for the entire distance; but the command was very fortunate, but one man, Henry E. Hard, bugler of Captain's company, being shot, the ball passing through the body, leaving an ugly wound, but at this writing he is doing well, with a fair prospect of recovery. Fourteen horses were also killed and wounded from all the companies engaged.

After driving them through the whole village, the General ordered the lodges burned and the captured stock to be driven up the sides of the canyons to the prairie above. While executing this order, some little sharp skirmishing occurred with straggling Indians, who had secured places among the rocks high up the sides of the canyon, which they had reached by trail that could not be found by our men, and from which they delivered fire with comparative impunity.

During the fight five Indians were certainly killed, and probably ten or twelve were wounded.

Fourteen hundred and six horses and mules were captured, from which 360 of the best were selected to be retained, and 1046 were shot to prevent the possibility of their again falling into the

The Long March North

hands of hostile Indians . . . the loss of this particular band of Indians is almost irreparable. Taken all in all, this is believed to be the most effective blow dealt the Comanches and Kiowas on this frontier during the last two years. Some of the servants and Tonkawa squaws found time to do considerable plundering in the camp while the troops were at the front fighting, and before it burned, and the plunder they got is worth mentioning. There are bows and arrows, shields, robes with hair and robes without hair, curiously decorated and painted, new blankets . . . stone china, kettles, tools — the best breech-loading arms, with plenty of metallic cartridges, one mule being found packed with 500 rounds and another loaded with lead and powder in kegs, bales of calico and turkey red, sacks of Minneapolis and Osage Mission flour, groceries of all kinds in profusion. Indeed, they seem to be richer in everything than white men who behave themselves.

The Seminole Negroes of Fort Clark, as usual, led the way. Mackenzie himself would report of the attack:

. . . the 1st Battalion was ordered down into the canyon 'A' and 'E' Comps 4th Cavalry with the scouts under Lt. Thompson had a running fight with the Indians for about four miles in which they killed three Indians and captured 1424 head of stock consisting of ponies, colts & mules. The Indians were all killed by the scouts under Lt. Thompson who was in the advance . . .

It had been a deadly day.

Captain Carter always remembered that the Indian gunfire was "hot and galling." Acting Assistant Surgeon Rufus Choate, running quickly to the fallen trumpeter, said later that the trooper had survived his wounds simply because he had fasted for thirty hours before the charge into the depths of Palo Duro Canyon.

Mackenzie saw a group of soldiers from Troop "H" dismount, and he called to the sergeant, "Where are you going with those men?"

"To clear the bluff, sir."

"By whose orders?"

"Captain Gunther's."

Mackenzie shook his head. "Take those men back to their company," he shouted. "Not one of them would live to reach the top."

Moments later, the colonel discovered Private McGowan kneeling over the horse that had been shot out from beneath him. "Get away from there or you will be hit," he yelled.

"Yes, sir."

But McGowan didn't move. He kept digging with his hands beneath the fallen horse as bullets tore up the ground around him.

"I told you to go away from there," Mackenzie repeated. "Are you going?"

"Damned if I am until I get my tobacco and ammunition," the private answered, and he continued digging until he freed the saddle pockets that had been trapped under the dying horse.

As the day wore on, the gunfire worsened. And soldiers found themselves pinned down, virtually trapped, at the base of those steep, rugged walls of Palo Duro. For some, their plight suddenly seemed hopeless. They had come a long way to die.

One asked, "How will we ever get out of here?"

Mackenzie answered calmly, "I brought you in. I will take you out."

And he did.

On that hot and galling day upon the Llano Estacado, he forever broke the spirit of the Plains Indians who pillaged the Red River country of Texas. The casualties had been few, but the destruction was great. Their weapons had been lost. Their horses were gone, and so was their will to fight. With horses, they had been a magnificent, a formidable fighting force. Without them, the Indians were as cripples. Most began the long, slow trek back to the reservation at Fort Sill, defeated and destitute. They were, at last, being forced to walk "the white man's road."

Captain Carter would write:

> During the four months the 'Southern Column' was in the field, the efforts to punish and subdue or exterminate the four hostile tribes . . . were unceasing. The Comanches, Kiowas, Southern Cheyennes and Arap[a]hoes had banded together and foolishly imagined they could whip out all the whites in that section of the country, but sad experience taught them that they were playing with fire. They did not recover for many a year.

General Sheridan expressed his belief that "this campaign was not only comprehensive, but was the most successful of any Indian Campaign in this country since its settlement by the whites."

During the Fourth Cavalry's march across the Staked Plains, Private Adam Paine became the second black and the first Seminole Negro to be cited for "gallantry in action." He had stood firm "when attacked by a largely superior party of Indians." And when Colonel Mackenzie recommended him for the badge of valor, he wrote that "this man is a scout of great courage." He would later say that Adam Paine "has, I believe, more cool daring than any scout I have ever known."

7

The Path of the Whirlwind

On the battlefield, the Seminole Negroes were a respected force in the American arsenal. Back home, down on the bleak fringes of the Trans-Pecos, the scouts faced life as outcasts, unwanted and exiled away from the mainstream of society. They were half-breeds and the prejudiced had no use for them. It was all right for the Seminole Negroes to sacrifice themselves on the prairie, but no one wanted them for neighbors. The government, still, refused to give them their promised land. It was as though officials believed that, sooner or later, the scouts would be wiped out one by one in the Indian wars, and no one would have to bother with them anymore. Bullets would relieve their diplomatic headaches and obligations.

The Seminole Negroes were generally orderly and law abiding, but trouble stalked their every footstep. Around a bottle of whiskey, late one night in Eagle Pass, the notorious King Fisher and one of the scouts argued, then fought. Gunfire suddenly erupted in the dimly-lit barroom. One bullet creased the outlaw leader's scalp and Corporal George Washington, Chief Horse's nephew, crumpled with a King Fisher slug in his gut. He would linger in misery for months before succumbing to the wound. Ill feelings festered like an infected boil through the streets of Eagle Pass as Christmas failed to bring peace on earth or goodwill toward men in 1874. The decision was reached to move those scouts who still resided at Fort Duncan on to Fort Clark. It was best to end a war before it started.

The outpost was glad the Seminole Negroes were coming their way.

Brackettville wasn't.

Some who lived in Kinney County flatly accused the scouts of sheltering horse thieves, and others spread hearsay that the Seminole Negroes were "constantly preying" on their property. The root of the conflict was land. Scouts had a tendency to go out and raise their crops on ground that belonged to someone else. Greed resorted to hired guns.

The shadows of twilight were hanging heavy over Fort Clark when Chief John Horse and Titus Payne walked around the corner of the post hospital. A gunshot out of the darkness killed Payne instantly, and another bullet ripped into the chieftain. He had just enough strength left to mount his horse, American, and escape the ambushers' guns. For weeks, the rumors blamed the assassination on King Fisher's men, but no one ever knew for sure. The gunmen were never found, never punished. Immediately, attempts were made to find the scouts a permanent home. But no one wanted to give up their land. No one wanted to sell it. Brigadier General E. O. C. Ord even suggested that the scouts be shipped to a reservation populated by hostile tribes. Perhaps the Seminole Negroes would be a "God-sent good influence" on the embittered, shackled warriors, he said. After all, the general had always been impressed with "their simple manners and religious tendency." No one paid any attention to Ord. The Seminole Negroes were left to wander the streets of Brackettville as the pariah of the prairies. Those alleyways were, at times, as dangerous, as venomous as the battlefields where the scouts had so distinguished themselves.

Adam Paine stood at a Seminole dance, just after midnight, when a sheriff slipped up behind him with a double-barreled shotgun. The scout was a wanted man. Reports said that he had knifed a Negro soldier down in Brownsville, and he must be brought to trial. The sheriff was a suspicious man who took no chances. Adam Paine was a warrior himself, cool and daring, courageous and maybe even desperate. He slowly raised his double-barreled shotgun and fired, so close that the blast set Paine's clothes ablaze. There was no time for him to defend himself. He never saw death coming. And the second black and the first Seminole Negro to be cited for "gallantry in action," slumped to the floor amidst the music of a New Year's morning.

Brackettville had become a sore in the sight of Fort Clark during the 1870s. Lawlessness walked the streets as though it owned them, and maybe it did. On the night of October 5, 1873, John P. Fries, the sheriff of Kinney County was, according to Laurence Hagarty, the presiding justice of the county, "fouly assassinated." A paper poster was promptly nailed to the buildings, offering a thousand dollars in reward money for the arrest and conviction of the assassins.

The town itself was no longer quite so isolated. In 1875 and 1876, troops from Fort Clark installed telegraph lines into the area. But it was still rowdy and rambunctious, where the only thing cheaper than a man's life was a woman's body. Civilization did not always bring with it civilized men.

An early traveler, Vinton Lee James, rode a stagecoach west from San Antonio and pulled into Brackettville for the night. He ate a meal of hot waffles and butter, then he walked out of the Sargent's Hotel and into the town's wild streets. James would write:

> ... you may imagine our surprise to find ourselves in the liveliest burg in West Texas, where the night life could only be compared to the saloons and gambling places that existed in the early days of the gold excitement of California and the Klondike. It was pay day at Fort Clark, adjacent to the town, where thousands of United States soldiers were stationed, and such an assortment of humans I never saw before. There were painted Indians with feathered head-dress, Lipan and Seminole scouts, and members of the famous Bullis band that was the terror to marauding Indians, with mexicans, white and negro soldiers, desperadoes, and other characters, all armed and ready for a fight or a frolic, with a sprinkling of fair females, soliciting for the bar, where several bartenders were as busy as ants serving liquid refreshments. Gambling devices of every description lined the floors of the saloon. Gold, silver, and greenbacks were in plain view on the tables, and the dealer shuffled cards for the many bettors who either lost or won, as lady luck would have it. String bands made music, while everybody was busy dancing, drinking, or gambling. Will was a natural born gambler, and immediately got in the game, only to lose several dollars before I could stop him. There were several of these sociable places and we visited all of them. One hall was owned by an old friend of mine, a bearded Irishman, who spoke with a distinct brogue, by the name of Tom Maloney, who formerly was a cowman. I knew him in 1874 when I went on a cattle gathering expedition to the Edwards Plateau beyond the head of the Nueces River. Maloney had charge of the remuda of cow ponies which were hidden away in the hills far from the ranch to be safe from Indian raids, as the Indians always raided the settlements. Maloney

made me stop gambling, saying that was his business, and advised me not to commence that bad habit. I knew his advice was good, but I had hard work to make my friend, Will, quit.

The revelry would not last forever.

Mackenzie — both in Mexico and on the Staked Plains — had broken the Indian stronghold on the southwestern prairie. No one, the army at Fort Clark knew, would ever be really safe until the last of those wild, nomadic renegades had been driven off Texas soil. They were no longer tribal warriors struggling to possess the land of their fathers. They were outlaws, fighting for food, fighting to survive, fighting in anger, fighting until there would be no one left to fire the final shot.

They crossed paths too many times with the man the Indians called Thunderbolt. Lieutenant John Lapham Bullis would become their shadow, forever trailing them, forever chasing them across Texas, never letting up, never letting them rest, keeping them on the move and riding for their lives until they had nowhere else to go.

In *Century Magazine*, the artist Frederick Remington wrote of him:

> He does not seem to fear the beetle-browed pack of murderers with whom he has to deal, for he has spent his life in command of Indian scouts and not only understands their character, but has gotten out of the habit of fearing anything. If the deeds of this officer had been on civilized battle fields instead of in silently leading a pack of savages over the desert wastes of the Rio Grande, they would have gotten him his niche in the Temple of Fame. But they are locked up in the gossip of the army mess-room, and end in the soldiers' matter-of-fact joke about how Bullis used to eat his provisions in the field, by opening a can a day from the pack, and, whether it was peaches or corned-beef, making it suffice. The Indians regard him as almost supernatural, and speak of the 'Whirlwind' with many grunts of admiration, as they narrate his wonderful achievements.

The Seminole Negroes worshipped him. Bullis was a friend and a leader. He traveled light and he traveled lean. He struck without warning.

It would be his assignment to rid the Southwestern prairies of those last ragged, scattered remnants of Indians, once proud and fierce and powerful, now merely fugitives who had lost but refused to quit or surrender. Bullis rode relentlessly after them and knew they could not evade him, knew they could not escape the uncanny tracking dexterity of his Seminole Negro scouts. He would pursue them to the ends of the earth if necessary, bringing calm to a troubled land.

The prey became the hunter.

From afar, the small bands of Comanche and Apache and Kickapoo watched him come. And they carefully concealed their tracks. They knew that they must eliminate John Lapham Bullis or their time in the West would be no more.

On an April afternoon in 1875, Lieutenant Bullis and three of his trusted scouts departed Fort Clark and headed into the desolation of the Pecos River country. They accompanied Company A of the Twenty-fifth Infantry as far as Beaver Lake, then marched west, running across a batch of Indian signs on the dry arroyo of Johnson's run. The trail, apparently left behind by the Comanches, appeared to consist of about seventy-five horses. Only a few had riders, but all, Bullis decided, were probably stolen. He, Sergeant John Ward, Trumpeter Isaac Payne, and Private Pompey Factor settled down for the night beside a watering hole in the rock. The grass was good, and their mounts needed it. They had marched fifty miles that day and, from the looks of the trail, there would be many rough miles more to go.

In his report to the Department of Texas on May 12, Lieutenant Bullis wrote:

> The following morning we left camp at 4:30 o'clock and marched southwest for about twenty miles and struck the Pecos at the mouth of Howard's Creek. At this place we went into camp and remained for about four hours. Then crossed to the west side of the river and marched about twelve miles west and into camp after dark without water, but had plenty of good grass for our horses; we marched this day forty miles and went over a very rough country. The following morning we left camp very early (3:30 o'clock) and marched west, toward the Rio Grande, for about forty miles; then marched southwest for about fifteen miles and went into camp, after dark, near the Rio Grande, without water, but had plenty of good grass for our horses; we marched this day about fifty-five miles and fortunately found plenty of water, standing in holes, for both men and horses."

The days, the hours, had been filled with monotony. The sun parched the earth before them. The water was never cool, but it chased their thirst for a time. Each day was different, yet each day was the same. Boredom overtook them, but it didn't whip them, although it seldom left their sides for very long.

On the morning of April 25, boredom fled. Bullis wrote:

> . . .we left camp at 4 o'clock and marched south for about eighteen miles and crossed the Pecos, about a mile above its mouth, at an Indian crossing; we then marched southeast for about six miles and went into camp at a spring, in a cave (known as Paint Cave). I would say here that in traversing the country between the Pecos and Rio Grande we did not see any fresh Indian signs, but plenty of old, nearly all of which went toward a shallow crossing of the Rio Grande, known as Eagle's Nest crossing. We left the spring at 1 o'clock p.m. and marched east for about three miles and struck a fresh trail going northwest toward Eagle's Nest Crossing. The trail was quite large and came from the direction of the settlements, and was made, I judge, by seventy-five head, or more, of horses. We immediately dismounted and tied out horses, and crept, back of a bush, up to within about seventy-five yards of them (all of which were dismounted, except one squaw) and gave them a volley which we followed up lively for about three-fourths of an hour, during which time we twice took their horses from them, and killed three Indians and wounded a fourth. We were at last compelled to give way, as they were about to get around us and cut us off from our horses. I regret to say that I lost mine with saddle and bridle, complete, and just saved my hair by jumping on my Sergeant's horse, back of him. The truth is, there were some twenty-five or thirty Indians in all, and mostly armed with Winchester guns, and they were too much for us. As to my men, Sergeant John Ward, Trumpeter Isaac Payne, and Private Pompey Factor, they are brave and trustworthy, and are each worthy of a medal, the former of which had a ball shot through his carbine sling, and the stock to his carbine shattered. Relative to the Indians, I would say that, in my opinion, they were Comanches, and were from Mexico.

The three scouts had reached their horses easily, had mounted, and were riding away when Sergeant Ward realized that Bullis must be in trouble. Glancing over his shoulder, he saw the lieutenant struggling with a frightened young horse, wild and only half-trained. The Comanches were swarming down upon him. Ward yelled to his compadres, "We can't leave the lieutenant, boys." Payne and Factor abruptly wheeled their horses around and struck back at the charging Indians while Sergeant Ward raced madly past the straining Bullis and scooped him up off the bullet-riddled ground.

John Lapham Bullis would thus be around to fight another day.

One month and three days later — on May 28, 1875, Sergeant John Ward, Trumpeter Isaac Payne, and Private Pompey Factor were given the Congressional Medal of Honor for their bravery that afternoon beside Eagle's Nest Crossing. Ward had been a farmer in

Mexico. Payne had enlisted in the scouts at the age of seventeen. Pompey Factor would desert the night that hired guns cut down Titus Payne and Chief Horse beside the post hospital at Fort Clark. But on that day in the Pecos River country, while the air smelled of cordite and the gunfire was deafening — and Lieutenant Bullis had turned to make his last stand — the scouts exemplified their daring and their contempt for danger in the face of possible death. They had risked their lives without a second thought. At the time, it seemed to be the natural and only thing to do.

Bullis had befriended them.

He had fought with them and for them.

There had been no thought of leaving him behind.

Even Lieutenant Bullis, after his narrow escape, was commended for his "energy, gallantry, and good judgement," upon his return to Fort Clark. It was said of him, "His own conduct as well as that of his men, is well worthy of imitation, and shows what an officer can do who means business."

Bullis always meant business.

He was so well respected by the residents of Kinney County, that in 1876, those citizens petitioned Major General Edward Otho Cresap Ord, then commanding the Department of Texas, and asked that Bullis be given a special detachment of seventy men so he could wage his own personal campaign of warfare against the marauding bands of Indians. The lieutenant and his Seminole Negro scouts were dispatched to join Lieutenant Colonel William R. Shafter further west.

On the afternoon of June 20, 1876, he and his scouts were returning to the outpost of Las Moras after a tiring and fruitless sojourn in the Davis Mountains. They ran across a fresh Indian trail that ran south and on into the Chihuahuan Desert of Mexico. Lieutenent Bullis paused, as always, beside the Rio Grande, staring for a time into a land where he had been forbidden to go. Mackenzie had gone, and he had been successful, and no one reprimanded him at all. For too long, the Rio Grande had been a barrier, an invisible wall that had sheltered renegades who deserved no shelter. The river, he decided, would be a hindrance no longer. Bullis waved his men forward, and they rode quietly across the mud and onto the sands and cacti of a foreign land. That ford would forever be known as Bullis's Crossing.

For three days, the Seminole Negro scouts searched for the roving band of Mescalero Apaches, pursuing them far into the mountains, scattering beyond the range of rifle fire. Bullis could inflict no damage on the thieves. He did recover thirty-six stolen horses and herd them again to American soil.

As the lieutenant camped that night, he realized that he had disobeyed the law he had vowed to uphold. He was obviously safe, alone in that bleak and barren country. No one would ever know of his escapade at all. But Bullis would never be able to outrun his conscience. Too, he feared that the angry Apaches might report him to the Mexican government. They, perhaps, might vehemently protest his invasion. At first, the lieutenant dispatched a courier to Fort Clark with information on his foray into Mexico. Later, Bullis felt that it was his duty and his responsibility to admit his violations in person. He instructed his scouts to return to the fort at a normal pace, driving the stolen horses along with them. Bullis saddled up and galloped away, traveling alone through some of the roughest, most harsh and unforgiving country that Texas had to offer. He rode for thirty-six hours, covering more than one hundred and forty miles, to make his report on the skirmish across the Mexican desert.

The government, perhaps, did not condone Bullis's decision, but he was not chastised, either.

Before June had ended, Lieutenant Bullis, leading ninety men, again slipped into Mexico, attacking a Lipan village near Saragossa, Coahuila. His raid was one of vengeance, even rage. During April and May, Chief Washo Labo and his Lipan warriors had struck a settlement in the rugged high country around Eagle Pass, murdering a dozen Texans and leaving them in the dust of a hot afternoon. The Indians, General Ord had said, would pay and pay dearly for their transgressions. Bullis found the village empty. Someone had given them warning, and they had fled long before the soldiers came. Bullis was disappointed, but not defeated.

He waited.

He was not an impatient man.

On July 29, as a summer sun baked the Mexican earth around him, Lieutenant Bullis led twenty Seminole Negro scouts and twenty Negro soldiers of the Twenty-fourth Infantry back across the Rio Grande. Their move had been made in secrecy and had been designed to strike quickly and decisively. For twenty-five hours, the scouts followed a trail that took them to the banks of the San Antonio River, near Saragossa. In the early morning light, Bullis saw twenty-three lodges rising up off the simmering desert floor. Crawling quietly through the shadows of twilight, the soldiers attacked while the Indians lay sleeping, firing volley after volley as the renegades, in desperation and self-defense, charged the thin army line. The battle was filled with chaos and confusion, it was ragged and

The Path of the Whirlwind

unorganized, a clash of hand-to-hand and rifle-to-club combat that stormed across the tattered village for a quarter of an hour. At times the soldiers and scouts were almost overrun and driven away, but no one retreated and no one gave ground.

Fourteen braves lay dying on the banks of the San Antonio. Four squaws, along with a hundred horses and mules, were rounded up and captured by the troops.

Bullis left the village in ashes.

Three of his men had been wounded, but none fatally.

The lieutenant turned back toward the Rio Grande, unaware that his toughest battle lay before him. He was heading home, back to Fort Clark, as the victor.

There was trouble behind him.

A much superior force of regular Mexican Army troops pursued him across the prairie, closing rapidly, determined to cut off his retreat and prevent his feet from ever again touching American soil. Bullis would have been no match for them. His men were weary and bloodied as the Mexican force rode down hard upon them.

As the lieutenant turned again to fight, Colonel Shafter and three hundred reinforcements sped to his support. The Mexicans, who had been eager for a fight, fell back as the ranks of the invaders suddenly swelled. For a time, the two armies stared at each other across the field of mesquite and chaparral, not quite ready for a confrontation, not quite ready to die.

Shafter and the Mexican officers argued and negotiated. Their words were arrogant and angry, their faces sweaty and sullen. At last, the Mexican Army turned away.

Bullis, with his scouts, his soldiers, his captives, and his stolen horses and mules, waded again into the muddy, yet cleansing waters of the Rio Grande.

By the summer of 1877, General Ord in San Antonio, ignored the animosity of officials in Mexico, and wrote a directive urging cooperation between the two nations. However, the order said,

> . . . in case the lawless incursions continue he (Ord) will be at liberty, in the use of his own discretion, when in pursuit of a band of marauders and when his troops are either in sight of them or upon a fresh trail, to follow them across the Rio Grande, and to overtake and punish them, as well as retake stolen property taken from our citizens and found in their hands on the Mexican side of the line.

For the Seminole Negro scouts, all trails were fresh, no matter how old.

The Rio Grande would no longer be a barrier to Bullis.

In the autumn of that year, he chased one small party of Indians into Mexico, trapping them in a short, heated battle and killing five of the braves. A month later, he was back in Mexico, riding down upon a pack of Apaches, with stolen horses, in a timbered canyon. But the Indians were too well entrenched in the land they knew best, and Bullis had too few men to fight the odds. He retreated to the Rio Grande, only to return seven days later with a stronger detachment to attack again. Accompanied by the Eighth Cavalry, under Captain S. B. M. Young, Bullis and his scouts stalked the Apaches far into Mexico, back into the foothills of the Carmen Mountains. From the high ground, strong and invincible, the Indians would make their stand. Days earlier, they had caught Bullis on a narrow ledge and almost destroyed him. They would try again. However, their advantage withered away in the face of a hundred troops whose gunfire battered the canyon walls, driving the Apaches farther from the village, now ablaze and in ruin. A warrior fell, then the chief himself tumbled down the mountainside clawing at the bullet wound that would take his life. The Indians broke and ran, leaving their stolen loot behind them. They had stopped Bullis for awhile, but not forever. The soldiers recovered seventeen horses, five mules, one burro, saddles, water kegs, fifty-four deer and buffalo hides, and large quantities of venison and horse meat. For the Indians, the approaching winter would be a long one, maybe even a cold one, with no home and no horses and no food to satisfy their empty bellies. Bullis's message was clear to them. The raids into Texas slackened. The Indians knew that they no longer had any refuge to protect them from the wrath of the "Whirlwind."

In June of 1878, Captain S. M. B. Young, with Colonel Ranald Mackenzie, took a small unit and departed a camp on Devil's River in pursuit of Indian marauders who had vanished beyond the Rio Grande. Following behind him was a stronger column of infantry, cavalry, and artillery, led by Colonel Shafter. He had one mission in mind, and that was to frighten the Mexican government, remind it to never again trouble a small, weak Cavalry strike force trying to return to its own homeland. For Shafter, it would be a display of might and of power that the Mexican Army could not disregard nor forget.

In his journal, Captain John S. McNaught wrote:

> June 17. Started at about 7:15 A.M. Arrived at a point on the San Diego River, where we halted to rest, at 8:45 A.M. Courier from General Mackenzie Camp arrived with information that he had made a Dry Camp at about 16 miles from this point . . . Received orders to have one day's ration cooked and ready to move

The Path of the Whirlwind

at 8 A.M. to-morrow. Mackenzie with 6 troops of Cavalry came in During the night. Commenced raining about 10 P.M. and rained very hard nearly all night. Almost impossible to keep fires going to cook the day's rations, bake bread, & Baked bread in Dutch ovens has to have live coals with which to do it.

June 18 — No movement until further orders. Moved at 11 A.M. Gen'l Mackenzie in command — Marched to a water hole where we camped at about 3 P.M. . . .

June 19 — Moved at 2 A.M. The Cavalry some distance in advance. Marched to the San Rodriguez River where we bivouacked. This morning Gen'l Mackenzie's Cavalry Command was stopped at Remolino, by Gen'l Valdez's Mexican Army — who with about 200 troops had orders to resist our entry into the town. The Cavalry dismounted — two troops thrown out as skirmishers — when Mackenzie and Valdez had a parley. Mackenzie told Valdez that he proposed to move forward at 2 P.M. and if his troops were in the way he would open fire upon them, or words to that effect . . . Moved at 1:30 P.M. to Remolino where we prepared for action — Mexican skirmishers deployed to stop our advance . . . Our skirmishers advanced . . .

The Mexican general decided not to call Mackenzie's bluff. Mackenzie did not bluff.

McNaught concluded: "The Mexicans withdrew entirely, leaving us to pursue our course unmolested."

Within five days, the punitive expedition force had returned to Fort Clark. There would be no more trouble from the Mexican Army, only resentment.

January of 1879 draped the prairies with a damp, miserable chill. It was no place for either man or beast, yet a band of Lipan and Mescalero raiders were moving suspiciously out across the frontier of Texas. Lieutenant Bullis, along with thirty-nine Seminole Negro scouts, fifteen cavalrymen, three friendly Lipans, and a former Comanchero, was ordered to follow them.

No one trusted the Indians.

No one understood their motives or could decide what their destination might be.

For thirty-four days, Bullis and his men kept the wandering band in sight, facing cold winds that cut deep into the marrow of their bones, moving for days without fresh water, riding over a torturous terrain. The winter chill hurt. But thirst became the worst, the most deadly enemy of all. On the desert, it drives men mad. The whole Bullis party might have perished if Sergeant David Bowlegs had not displayed his knowledge of the land and its secrets, locating a sleeping spring and causing it to flow again. The chase didn't end until the renegades arrived at the warm, safe confines of the Fort Stanton Reservation in New Mexico. There had been no battles, no hostilities, just a silent march upon a prairie bitten by frost and freezing rains.

Bullis demanded that the fugitives be surrendered to him. He wanted to return them to the territory stricken by their depredations.

The agents at Fort Stanton refused to give them up.

Bullis, empty-handed, had no choice but to ride again toward Fort Clark. His long march would last for eighty days and cover 1,266 miles. For Bullis, it was an uncommon ride amidst uncommon circumstances. Few, if any leaders, it was written, could have kept their men together under such harsh, wretched conditions.

But Bullis wasn't an ordinary man.

And his scouts believed in him, would never forsake him.

Bullis had always been the one officer who had made them feel as valuable and as important as they were. He never ignored them or looked down on them. He was there whenever any of his men needed him, on the battlefield or at home.

Anytime a child was born in the Seminole Negro camp at Fort Clark, John Lapham Bullis would always appear within two or three days to inspect the infant. If it were a male, the lieutenant would grin broadly, raise the child triumphantly into the air, and say loudly, "This one'll make a mighty fine scout someday." The mother would smile. The father would burst with pride. His loyalty for Bullis would be thicker than either blood or water.

Bullis cared.

And the Seminole Negro would fight for him.

In the spring of 1881, they launched their last major campaign against a hostile band of Indians. The renegades, for the most part, had turned their backs on Texas. It held the bones of too many of their dead. The land had become worthless to them. Only with whiskey in their bellies did any of them ever return, and sometimes their rides bordered on insanity.

On April 14, a drunken band of Lipans stormed down on the home of Mrs. McLauren near the head of the Frio River, killing both the woman and a young boy named Allen Reiss. The renegades quickly looted the house, packing away everything they could find of value before riding away.

But they respected the Seminole Negro scouts.

They feared them.

They knew that, sooner or later, Bullis and the scouts would be coming after them.

So the Indians killed a horse, tore off its hide, and wrapped the blood-drenched leather around the hooves of their horses. They would leave no tracks. They would leave no trail for the Seminole Negroes to follow.

Bullis didn't learn of the carnage until twelve days after the attack. But he received a pursuit order from General Stanley that instructed him "to pursue and destroy." At first, it seemed to be a hopeless task. But the keen eyes of his scouts uncovered the trail, even though it had been virtually obliterated by the horse-hide padding.

They would not lose it.

The country south along the Devil's River was rough and rocky, seamed with arroyos, and swept by the winds. The scouts steadily and relentlessly followed the cold tracks until they crossed the Rio Grande about ten miles north of the Pecos. Bullis camped on the night of April 30 on American soil and quietly outlined his plan of attack. It would be one of surprise. It would be one of destruction.

That was the only way John Lapham Bullis knew how to fight.

As a red shade of dawn lightened the sky, the lieutenant and his scouts slipped again into Mexico, working their way back into the Sierro del Burro Mountains. It wasn't until four o'clock the next afternoon, that a Seminole Negro discovered the raiding party lying "in a rough and broken country." The Lipans, careless in the belief that none could trail them, were bedding down for the night.

Bullis let them sleep.

At midnight, he carefully and quietly moved twenty-seven of his scouts into position. They lay, shapeless forms against the rocks, and waited for the dim light of morning to crawl again across the desert.

At daybreak, Bullis struck.

Four braves died before they had time to react, unaware that danger had hovered above them like a vulture among the rocks. A sub-chief of the Lipans, San-da-ve, fled wounded. It would be learned from the Mexicans who found him, that the renegade leader had not survived the brief and deadly battle. Bullis rounded up a woman and a boy, as well as twenty-one animals. Mrs. McLauren and Allen Reiss could, at last, rest in peace. Their deaths had not gone unpunished.

The Indians abandoned Texas.

During the next year, twelve expeditions from a dozen outposts scoured the countryside, riding for 3,662 miles, and not one of them found any trace or any signs of Indians.

Texas, for a time, was at peace.

Bullis had made good his pledge to free the western frontier from renegade atrocities. For eight years, his Seminole Negro scouts from Fort Clark had fought bravely at his side, and none of them had ever been killed or even seriously wounded in battle. A descendant of the scouts would say, "The old people in those days were so loving with one another. That's why things went the way they did in the fighting; the old people were doing some powerful praying." Another believed, "When you are fighting for the right and have your trust in God, He will spread his hand over you."

In 1881, the residents of West Texas presented Lieutenant John Lapham Bullis with an engraved sword, and the inscription read: *He has protected our homes — our homes are open to him.* On April 7, 1882, the Indian fighter was honored by a joint resolution adopted by the Texas House and Senate for "the gallant and efficient services rendered by him and his command in behalf of the people of the frontier of this State, in repelling the depredations of Indians and other enemies of the frontier of Texas."

An old newspaper clipping said of him:

> He demanded no hardship or risk of his men, that he did not share, and due to his indomitable courage the scourge of murder and arson that reigned in Texas outpost settlements after the Civil War was ended.

It had been a turbulent era, full of turmoil and trials. Fort Clark remained unscathed, but the men who rode away from the post on patrols experienced the worst that the frontier had to offer. Indians waited for them. Sandstorms burnt them. An unrelenting sun penetrated the ground beneath them. Thirst nagged them. A merciless landscape taunted them. And many never left the ground that became their final resting place.

Upon a tombstone in Fort Clark's graveyard there was an the epitaph of a soldier, that symbolized so many:

> *O pray for the soldier, you kindhearted stranger;*
> *He has roamed the prairie for many a year;*
> *He has kept the Comanches from your ranches —*
> *And followed them far over the Texas frontier.*

8

Battle of the Bloody Bend

Fort Clark had long been a source of consternation and embarrassment to the United States Government. The Army had invested thousands of dollars into establishing the outpost on Las Moras Creek, but it didn't own the ground beneath those stone buildings. Mary Maverick did. For years, the government had been paying her rent on the 3,866 acres of land that held the men, the quarters, the horses, and the mules of the cavalry who were waging such a critical war against Indian raiders who scourged the prairies from Mexico. By June 30, 1878, the rental had risen dramatically from $600 a year to $1,200 a year.

The Army was in a quandary. Some military men wanted to buy Fort Clark. Others wanted to abandon it. Some argued that Fort Clark was absolutely necessary until, at least, the Comanche, Kickapoo, and Lipan warriors had been annihilated or chased from Texas soil. Others voiced that opinion that the Indian threat had greatly diminished and had probably been eliminated. To them, the fort was merely a lonely, isolated outpost on the edge of the desert that had lost its reason to survive.

As early as 1873, a board of officers was appointed to consider the subject of acquiring sites for permanent military posts throughout the country. And the board reported favorably to the Secretary of War, recommending that 700 acres of Fort Clark be bought for ten dollars an acre. At the time, it seemed to them like a good bargain. But Congress never did get around to appropriating any funds for the purchase, and the army kept doling out rent to Mary Maverick.

In May of 1878, she finally offered to transfer the entire leased site — all 3,867 $^1/_2$ acres — to the government for $25,000. Lieutenant General Philip Sheridan promptly dispatched a letter to the Adjutant General saying,

> Not one moment should be lost in securing the offer made by the owner of the site of Fort Clark. The post will have to be retained for many years, it is of great value to us in every way, and the price asked is very low.

Again the purchase was recommended by the board. Again Congress failed to find any money.

Maybe the fort had indeed lost its value, some reasoned. After all, in February, Lieutenant H. W. Lawton had written that the stables and the quartermaster storehouse — contemplated in the original construction of the fort — had never even been built. He pointed out that they were however,

> ...absolutely necessary for the proper protection of the animals and public property, much of the latter being now out of doors subject to deterioration from action of the weather and loss from other causes.

The lieutenant added that the post badly needed "pointing up, repainting, new floors, repair of roofs, etc., etc." He was concerned because arrangements had never been made to bring water into the garrison as a protection against fire if nothing else. He wrote, "... two sets of officers quarters have burned down, either of which cost more than the machinery for putting the water up." Lawton determined that it would cost thirty thousand dollars to finally complete and thoroughly repair Fort Clark the way it needed to be refurbished.

By August, it appeared that General Sheridan had indeed been correct when he wrote that Mary Maverick's asking price of $25,000 "is very low." She had had second thoughts. Perhaps she was merely weary of bureaucracy moving in such slow and mysterious ways and sometimes not moving at all. Toward the end of summer, Mary Maverick raised the price of the land to $50,000.

It wasn't until April 16, 1880, that Congress, at last, got around to appropriating $200,000 for acquiring sites and erecting posts that would protect the length and breadth of the Rio Grande frontier.

Battle of the Bloody Bend

Fort Clark, it seemed, would finally become government property after all.

But Mary Maverick withdrew her offer.

The land wasn't for sale at all.

By 1882, the fate of Fort Clark had been placed in nervous, uncertain hands. Early in the year, General William T. Sherman, under orders from the Secretary of War, inspected the military posts whose guns stood guard over the Land of the Rio Grande. He definitely wasn't impressed with Fort Clark, describing it as the "largest and most costly military post in Texas if not in the United States." And, according to his rough estimations, a quarter of a million dollars had already been spent on the garrison during its three turbulent decades. Most of that, he decided, had been wasted. Fort Clark, as far as Sherman was concerned, was useless, made obsolete by the construction of a new railroad nine miles to the south. Fort Clark had long been isolated. Now, it should be forgotten. Sherman much preferred that Fort Duncan be maintained and expanded instead.

Fort Clark stood on the threshold of abandonment.

In Brackettville, the "citizens, cattle and sheep raisers, merchants, property owners, tax payers, and voters" of Kinney County all got together in desperation and petitioned the Honorable S. B. Maxey, their senator.

They told him that if the military marched its soldiers away from the post, "it would take away the protection afforded us against Mexicans, out-laws, cattle thieves, and renegades from justice." Because of the fort, those people said, "we now have a peaceful and law abiding community and country." Too, the presence of the post in the vicinity, made their property worth at least five million dollars. Should Fort Clark become darkened and deserted, they argued, their homes and their property would be virtually worthless.

> ... such abandonment would stop further settlement and development of this and adjoining counties, and would very soon depopulate the same and bring ruin to this entire section of country by leaving it unprotected and worthless.
>
> We therefore pray you, for the interests of the Rio Grande frontier, settlers, stock raisers and others, to use your most earnest and valuable efforts to urge upon the Secretary of War to avert the ruin and disaster such abandonment would bring upon us, and the great and indispensable benefit and absolute necessity the said Post is for the safety and welfare of this entire border.

A year later, the Army received its annual inspection report of Fort Clark:

> Some of the barracks are rather small and poorly built. Kitchens and mess-rooms are detached temporary structures . . . There is ample supply of good water, well distributed through the post . . . New cemetery recently enclosed with a good wooden fence. Daily schools for children and enlisted men. Facilities for gardens limited and poor. Gardens were made by three companies with varied success and failure . . . Officers quarters are insufficient for present command, some additional quarters should be built. Also, a school house, chapel, library and quarters for non-commissioned staff should be built.

The anxious residents of Kinney County waited — with some fear and trepidation — to see if the Secretary of War would listen compassionately to their pleas in behalf of Fort Clark or whether he would be influenced by the troubles and shortcomings of the post that had been magnified in the report.

General Sheridan reiterated his support for the fort, reporting, "It is a good post for either cavalry or infantry, or both; buildings worth about $500,000."

The General of the Army, however, chose a temporary compromise that appeased no one.

He wrote:

> As San Antonio has been chosen by general consent, and Fort Clark is on ground belonging to Mrs. Maverick, who asks an enormous price, we are forced by recent laws to purchase or abandon. After careful personal inspection, I am convinced that in the long run it will be more economical to use Fort Clark for a few years under existing lease and then abandon it for San Antonio.

Some, however, didn't give up hope.

Citizens were fighting for Fort Clark. So were some military men.

As the autumn of 1883 rode chilled winds across the prairie of Texas, General Ranald Mackenzie sat down and began his own personal negotiations with the Mavericks for the purchase of Fort Clark. He promptly wrote Sheridan, telling him confidentially,

> I can now buy the entire track now under lease . . . for sixty thousand ($60,000) dollars . . . I regard this as a very reasonable purchase at this time and as of the first importance — I do not see how we can get along and give Troops shelter without this post . . . At this time the post can be bought simply on your recommendation and the approval of the Secretary of War . . . If this transaction is authorized please furnish me the necessary authority to make it —In the negotiations which were conducted with the

Mavericks the understanding was that — I had no power of committing the government but simply to recommend. I do not think we should bother about the rise in price since the last offer as the rise in property generally on the frontier has been great, but close at once with the offer. The Mavericks have been very courteous and liberal in my conversation with them . . . this property cannot in my judgment hereafter be bought at this price.

A month later, Major General J. M. Schofield reported that Ranald Mackenzie was suffering from "mental aberration" and was not able to even exercise his command. He would be placed in the Bloomingdale Asylum in New York, a tragic ending for a man who had given so much of himself to tame a frontier and prepare a way for the footsteps of civilization to follow him.

His conversations with the Mavericks obviously ceased. Fort Clark had lost one of its strongest, most supportive voices. The Army should have listened to Mackenzie during those last days when he fought to maintain a grip on his sanity. Mackenzie had been right in his belief that the property could never again be bought for $60,000.

A year later, the Secretary of War finally dipped into his special $200,000 Rio Grande appropriation and came up with enough money to purchase the land that lay beneath Fort Clark.

It cost him $80,000.

Sherman wasn't pleased. Sheridan was. In his report, he had gone out of his way to glorify the finer points of the old Texas outpost, writing:

> The water supply is scarcely surpassed, in fact is unequalled at any point in Western Texas. The post presented the appearance of a well built village, the material being stone . . . The officers' quarters are well built and commodious. The men's quarters are roomy . . . The storehouses are good, especially the quartermasters storehouse, which is the best and most complete in all its appointments of any I have ever seen in the army.

To Sheridan, the $80,000 was a fair price that no one should complain about. He wrote, "I doubt if we can erect in any part of Texas a similar large post for less than $300,000."

Brackettville could rest easy for a change.

However, it remained a rowdy and somewhat reckless little hamlet upon the prairie. Cowboys trudged its streets, drank the whiskey from its saloons, and slept in its alleyways. One said it made him feel like the frazzled end of a misspent life. And another swore, after a hard night with a cold bottle, "I had a taste in my mouth like

I'd had supper with a kiyote. If I'd had store-bought teeth, I'd a taken 'em out back and buried 'em." One cowhand, after too many days on too many trails, described his whiskey as "a brand o' booze that a man could git drunk on and be shot through the brain and it wouldn't kill 'im till he sobered up."

Brackettville became, for a time, a den of iniquity and proud of it. In June of 1881, the county court records revealed criminal cases involving a disorderly house, fornication, adultery, exhibiting monte, selling intoxicating liquors, and unlawfully cutting timber on land not his own. Teamsters rolled into town, trying to leave and trying to forget the sun-shrunken prairies that shimmered just beyond the last shade of mesquite. Investors and merchants slipped in from San Antonio. Buffalo hunters staggered in from the dusty plains of Kansas. Professional gamblers just dealt from the bottom of another deck of cards. The back alleys of Brackettville were home for claim jumpers and cattle rustlers and old soldiers. It was sometimes difficult to tell the outlaws from the law.

King Fisher came to town with hired guns. The son of Colonel William B. Travis, who fell at the Alamo, was a member of the Texas Rangers stationed at Fort Clark. Wyatt Earp spent many of his winter months in the mild climate of Brackettville. They all believed that God, perhaps, created all men. But it was Colonel Colt with his .45 caliber pistols who made them all equal. If someone died from a gunshot, it was generally because he was a thief, an outlaw, or just a little slow.

In the distance, a division of the Southern Pacific Railroad known as the Galveston, Harrisburg, and San Antonio Railroad was frantically racing the International and Great Northern Railroad toward Mexico. Originally, the tracks were planned to cut through Brackettville itself and become a smoking lifeline to the skylines of civilization. However, the town snubbed the rail line, and Southern Pacific officials decided that they didn't really need Brackettville after all. Some insist that the city leaders were too confident. The train was bound to come rolling past Fort Clark. So they didn't even bother to raise the matching amount of money that the railway officials required. Others argue that the city leaders just didn't want a railroad interfering with their town. For too long, men had depended on government contracts to haul freight from the fort, and they weren't about to let a train come in and take that money away from them.

The Southern Pacific went elsewhere.

Its rails turned southward for ten miles to a little crossroads

Battle of the Bloody Bend

that would become known as Spofford Junction. A branch line angled on down to Eagle Pass.

Brackettville had damned itself.

The Seminole Negroes felt that they had been damned for years. Even though the Indian wars had fled westward, they still clung like children to the outskirts of Fort Clark. The post kept them fed and sometimes clothed, and it was their security upon a land that had none. When Walter Harper rode into Brackettville, he watched the old scouts closely, even built a small grass-thatched hut near their encampment around the fort. Harper wrote of them:

> They were a reckless, hard riding, hard fighting, fearless bunch. They spread terror to the Comanches wherever they encountered them. They were a turbulent noisy lot. Their language among themselves was a jargon of Spanish, English and Seminole, but they used very good English or Spanish when speaking to outsiders. They seemed to be of a religious turn, holding frequent services which consisted mainly of exhortations by one or more members, accompanied by a kind of groaning chant from the rest of the congregation, together with frequent screams and shouts. The excitement gradually increased until it became a bedlam of noise. These meetings began about dark and lasted until eleven or twelve o'clock.
>
> They seemed to be a strictly honest bunch. My friends told me they had never been known to molest anything not belonging to them. They cultivated small gardens which were irrigated from the Las Moras Creek. These gardens were cultivated by the women and older children. The ground was dug with a hoe and the seeds were planted haphazard, without regard to rows or distance. Crops consisted mainly of corn and melons of various kinds. The bucks, when not on duty, stayed at home with their families but were subject to call at any minute.

For Lieutenant Bullis, the condition of the Seminole Negroes had become pitiful. There were, he reported, about 225 members of the tribe hanging around Fort Clark "all of which are destitute," except those who had enlisted. Bullis wrote, "I would say that old and young can be seen every day picking food from the slop barrels." The commanding officer of the post habitually issued rations to the most needy, the fatherless children, the aging men and women. Too often, it wasn't enough. The Seminole Negroes simply ate what they could, when they could, and they waited for land that had been promised but not given them. It would be a long and futile wait.

When Jasper Ewing Brady arrived at Fort Clark in 1889, he discovered that civilization had forsaken Brackettville. He later wrote in the *New York Herald Tribune*:

> The population was some 500, mostly Mexicans, Saloons were on every corner and plenty in between. Dance halls, brothels — let your imagination run riot and you may know approximate what this town was in those hectic days. Two dance halls stood out — the Blue Goose and the Gray Mule. There were several kinds of dances indulged in that are not seen on stage or ballroom floor. There was cheap liquor, cards, all kinds of gambling, women and no legal restraint . . .
>
> The first day after our arrival at Clark will linger in my memory as long as I shall live. It was wild — not a revel, but an orgy, such as would have made Nero look like a piker. Mrs. Boole in her most vivid dreams could never conjure up such a scene. It was uncontrolled liquor at its worst. Mind you, there was no legal restraint of any kind in Brackettville. The so-called citizens made their living from it.
>
> The morning after pay day there were 110 men confined in the guard house or under arrest charged with every crime in the calendar, from drunkenness and A.W.O.L. up to and including attempted murder. The next pay day was a repetition, only there were 112 men in hock.
>
> Then came a sudden change — the post trader's stores were abolished and the army canteen came into being.
>
> No liquors were sold — hard ones, I mean — only beer and wine so light it would not stand alone. Men were allowed credit up to a certain small percentage of their pay. No man could buy more than 5 glasses of beer a day.

Then a change came over the face of the garrison. Brady explained:

> The canteen had been in operation about a month when along came a pay day. Here was the test and here is the answer. The day following found 8 men out of 1,100 confined in the guardhouse. The next pay day there were 10, and the following one there were 6. Never was there over a dozen. It was the finest example of controlled temperance I have ever seen, and everything was working against that control. Those figures are real, I was doing clerical work at the time and know.

The prairie was, for a time, at peace with itself.

Even Brackettville was settling down. C. A. Windus, the proprietor of the Coffin Stage Line introduced a new Concord Stage that would haul passengers from the Spofford railhead to town for a dollar apiece. A scaffold, to be used for hangings, was raised in the county jail. The government began shipping cement from Houston for the construction of buildings at Fort Clark. By 1888, milk was selling for ten cents a quart. Sheared muttons were bringing $2.25 a pound. The Hilton House was advertising board and rooms for eighteen dollars a month — five dollars a month if you only wanted a nice furnished room. William Locke, the town jailer, was angry because someone had rustled his "best and fattest" steer down on the Rio Grande.

During the next decade, in downtown Brackettville, citizens had their chests checked by Tom's lung tester at the Favorite. C. A. Windus decided to run for county judge, but he quickly pointed out "the expenditure of money for the purpose of gaining votes is not in my line." Isabella McKerrow offered a thousand dollar reward for "the person or persons who waylaid and shot my husband, J. R. McKerrow." And when the Spanish–American war broke out, the newspaper lamented, "Seems like since our boys left for Manila the young folks in town lost all interest in dances, entertainment, etc."

The Roach and Company was selling everything from cradles to coffins, at reasonable prices, of course. Doctor Mary Walker was seen walking the streets in men's clothing, which was almost a disgrace. Mary Walker had been a surgeon during the War Between the States and was the only woman to ever win the Congressional Medal of Honor. However, in 1917, she would be one of 911 whose medal was taken away by the Army Medal of Honor Board.

The *Brackett News* urged citizens:

> Now is the time to show your grit. Talk up your town and county, help advertise the resources; tell strangers what a fine town we have. Don't for goodness sakes squat around lamenting and talking hard times. Get to work.

The *News* also reported the story below, which is an exact quote. Perhaps the reporter — or the typesetter — had visited the saloon before doing the story:

> An exciting runaway occurred on Main Street Tuesday evening and for a minute there was a wild rush for a place of safety. John Gilder had left his horse and cart near Roach & drop in for a drink at Fritter's. He have become frightened at his shadow, for he kicked at it once and then lit out. He started up the middle of the street alright, but an uncontrollable desire seized him to smash something for Nance and Struder & drop in for a drink at Fritter's. He tore the casing off Nance's door, missed taking Bobby Nance's leg off by a scratch, swung a shutter loose and busted the mourner's bench at Studer's scraped the flagpole of liberty at Fritter's door and broke away from the cart under the gallery. The cart and horse were uninjured but the harness was badly pied. August was

feeling a bit sick anyway and the excitement aggravated the malady. Nance had a severe attack of nervous prostration but is recovering.

Fort Clark, by the late 1880s, had become a melting pot for soldiers of many nationalities. Private Michael Brown, for example, had Russian ancestry. Private Wilhelm Schouburgh was German; Private Pierre Gerard, French; Private Joseph Lafani, Italian; Private Adolph Kurg, Swiss; Private Audeos Pedersen, Denmark; Private George Dine, Canadian, and Private Peter Moborg, Sweden. And, in addition, there were also troopers who had descended from Polish and Welsh lineage. Theirs was not an easy lot in life. Trouble, it sometimes seemed, dogged every soldier's footsteps. So did loneliness and homesickness and boredom. The *Brackett News*, in 1888, reported that, "Private Daniels, Company 'D' of the 23rd Infantry, and Private Jones of Company 'E' have deserted, and Private Loomis, of Company 'H' 23rd Infantry, was relieved from guard and placed in confinement a few nights ago. He hid his gun, belts, etc. while on post and went to town. While absent his gun was stolen. He was then, plainly speaking, 'out of luck.' "

At Fort Clark, disease became the toughest fight of all. The Texas prairie tested men, and the weak ones sometimes perished.

Colonel James P. Kimball, while at Fort Clark, found it necessary to be a surgeon and a dentist, as well as a general practitioner. He treated about as many traders and travelers and ranchers as he did soldiers. His wife recalled Colonel Kimball's first notable case on Las Moras Creek. She wrote:

> A soldier had fallen asleep in the sun, and during his unconscious half-hour a fly peculiar to the tropics deposited its larvae in his nostrils. It was only heroic treatment with chloroform that destroyed the maggots and saved the man from suffocation.

It was, however, "Texas Fever" that took a tragic toll at the fort. The disease gripped the outpost and silently stalked the soldiers who had been assigned, or cursed, with duty on Las Moras. It raged like the plague through Fort Clark, and no one, it seemed was immune to the lethal fever. Even Dr. Kimball became one of its victims. The malady, racing out of control, was finally identified as typhoid, caused by a contaminated water supply.

Mrs. Kimball, watching over her fevered husband both day and night, remembered,

> Daily the Hospital Steward reported new cases, and repeatedly

the funeral march sounded across the parade; each time I trembled lest it should sound again in front of our door. In those days, the Red Cross army nurse was unknown, and I, who had little knowledge of sickness, was obliged to learn nursing under the tutelage of my patient.

The doctor recovered, promptly took his sick leave, and put Fort Clark far behind him.

Fort Clark slowly pulled itself back together again as the sounds, the tremors of the Spanish–American war reached across Texas and rolled into the garrison on the Rio Grande. Teddy Roosevelt, who always knew how to walk softly and carry a big stick, was in the Menger Bar in San Antonio, recruiting his Rough Riders. By 1898, the Third Texas U.S. Volunteers Infantry found itself at Fort Clark, gearing itself for battle. They didn't consider it to be a glorious assignment. One chaplain would later write:

> To have served at Clark, was at one time nearly equivalent to honorable mention, for such an entry in one's record was a sure token that the fortunate or unfortunate individual had been really initiated into army life.

To some, duty at the post was akin to having been banished to the Foreign Legion. Names and faces tended to get lost, or at least misplaced, down among the live oaks and pecans of Las Moras.

By the turn of the century, a nickel publication was being printed at Fort Clark, called — *The Third U.S.V. Infantry Magazine*, and one issue reflected the cynical position of the troops who drilled at the outpost. The letter was entitled "Farewell Senors," and it was written from the Third Texas to the Spanish Generals:

> ... But the chief thing you should be glad of— though we are not — is that the 'Fighting Third' was sidetracked.
>
> We drilled in our shirtsleeves so to speak — because we had no uniform — we had blood in our eyes — and a vacuum in our pockets. To harden ourselves we slept on the wet ground with a cold norther blowing and had the sky for a roof— because we couldn't get tents. We drilled till our shoes were worn out— then went to bed because we had no more.
>
> How anxious we were to pit our 'Civil War relics' against your improved Mousers but we were sent to guard the frontier, the rear being well protected — every man having from three to six patches on. The Third was being held — we were informed — for the final coup de grace, when like Napoleon's guards of old we were to come thundering down upon you with our indomitable valor — and our old guns as clubs — aided by a few Comanche

yells, to sweep the island of all foe. But unfortunately for you the protocol was signed and the Third still held down the frontier.

And now, as if slighting is not sufficient, insult is added and the 'Fighting Third,' the valorous, the veterans of Texas — the kickers — the would-be history makers, without being given a chance to clean out anything — except the canteen — is to be saled down, or rather, mustered out.

If at any time you or any nation who can appreciate a regiment built strictly for business, call on the Third Infantry, but garrison duty on the Rio Grande, never!

For soldiers at Fort Clark, war had always been more formidable across the border than across an ocean.

The army just didn't trust Mexico.

Bitterness simmered from those days when roving bands of Indian renegades found safety and shelter on the southern landscape beyond the Rio Grande. Too many in power still believed that the United States lay in danger of hostile attacks from forces in Mexico.

The border must be protected.

Fort Clark would be The Sentinel.

Far to the West, the Big Bend country loomed upward as a bastion of rock and desert, impregnable, a natural fortress for bandits and outlaws who wanted to hide away in a land where few dared to tread and hardly anyone chose to live. When the residents of Pecos County asked for additional Texas Rangers to guard them, they wrote that the region "is a favorable resort of the murderers and desperadoes driven from other sections of the state." Robert T. Hill, who trekked the land around the turn of the century, called it the "Bloody Bend." And he referred to it as a place where "civilization finds it difficult to gain a foothold." Down among the Chisos Mountains, civilization wasn't even trying. With a 250-square-mile area, by 1910, only twenty ranches had been scattered around the fringe of the Chihuahuan Desert. A few Rangers rode the countryside. Fewer cavalrymen patrolled the bleak and barren stretch of land. The manpower and the firepower were simply too small to liberate the country from the bandits who left a searing brand of death and ruin upon the face of the Big Bend.

In Mexico, a revolution erupted in 1911, and the soldiers of Pancho Villa fanned out along both sides of the Rio Grande in search of food and supplies and ammunition. They didn't buy what they wanted. They took it. And they had no pity on anyone who stood in their way.

Texas did.

The revolutionary forces attacked Ojinaga, and frightened,

penniless Mexican refugees were strung out across the desert, stumbling northward, hoping to find safety at Marfa. They found a train that would ship them on to a crowded, miserly camp in El Paso that had been hastily erected to hold the runaways. A second raiding party struck the Lamar Davis Ranch and escaped with rifles, saddles, horses, mules, a handful of ammunition, and seventy bushels of wheat. The bandits looted both Boquillas and San Vicente on the Mexican sands of the Rio Grande.

For a time, Lieutenant James L. Collins and twenty-five men were dispatched to ride as sentries along the river. But in 1913, he sent word to J. O. Langford, who operated a bath concession at Hot Springs, that he and his troops had been ordered to leave. "That word shook me up," Langford would say. "I realized that Collins would not have risked court martial for revealing army secrets if he had not felt that our danger here was immediate and great." He and his family packed up and journeyed northward with the refugees. Langford could not fight the revolutionaries alone. A few men tried. Alpine sheriff, J. Allen Walton, sought to chase them down, but the vastness of the Big Bend swallowed them up, and the posse rode home empty-handed. As the *Alpine Avalanche* said of that desolate, broken country,

> It would be a hard matter to locate anyone familiar with that section who was trying to avoid detection. As the Mexicans are known to be bad men and are armed to the teeth, there has been much uneasiness felt about those who pursue them . . .

Pancho Villa's men plundered ranches and tiny settlements in the name of Mexican independence. Chico Cano's raiders looted the land because it was the easy road to ill-gotten wealth. Men fought. And some died. The plea went out from Texas Governor Oscar B. Colquitt to President William Howard Taft that troops were needed to safeguard the interests of those who called the Big Bend home. By 1915, Lindley M. Garrison, the Secretary of War, even admitted that the problem in West Texas was "very near to justifying martial law." He was obviously aware of the trouble, but he took no action to end it. After all, General Funston, the commander of the Southern Department, had told him that the situation was primarily a local one, a problem that the sheriff and a few Texas Rangers should be able to quell by themselves. Besides, the Secretary of War told President Taft,

> The distance from Brownsville to El Paso . . . is over twelve hundred miles, following the windings of the river, and part of it is

about as inaccessible and difficult a country as can be imagined, so that it will not in any circumstances be possible to station troops throughout that entire stretch.

So Villa's band ran rampant and often.

However, in March of 1916, a group of soldiers happened to be near when the guerrilla warriors tried to wreck a train just east of Alpine. And they captured the gang before the railroad could be harmed or damaged at all. At times, the military was in the right place at the right time. But mostly, it was just scattered and lost within a land as hostile as the revolutionaries themselves.

On the night of May 5, 1916, Glenn Springs lay sleeping. It was nothing more than a small general store, surrounded by a few mud huts, that C. D. Wood and W. K. Ellis had built to supply the Mexican peons who worked in their wax factory. In the mud huts slept a small contingent of troopers from the Fourteenth Cavalry, unconcerned and unprepared.

At midnight, eighty bandits struck, riding hard into the ragged village, screaming and yelling, firing wildly into every dwelling, trapping Sergeant Charles Smyth and six soldiers in a cramped adobe cook house. The fight wore on, and suddenly one of Villa's men pitched a flaming torch onto the thatched roof, and the hut crackled and roared with a fire so bright that it chased the darkness from Glenn Springs. The soldiers tumbled out of the blaze, and three were cut down by the bandits' bullets. The sergeant and his other three men fled into the night.

One revolutionary kicked the door in at C. G. Compton's home, and the men rushed inside. Compton had taken his young daughter to another hut and was returning for his seven-year-old son, Tommy, when a Bandit shot the small boy as he huddled alone in the house. Compton's other son, a deaf-mute, ambled confused and frightened amidst the gunfire and the fighting, emerging unharmed as the shooting died away.

In the general store, the bandits took everything they could pack and carry on horseback, leaving behind only the sauerkraut. The Mexicans hauled it away once, then smelled it, and threw the bag away, believing it spoiled.

Just before daylight, Jesse Deemer was abruptly awakened as a second arm of the revolutionary gang charged down upon his little store in Boquillas. Deemer was outmanned, and he knew it. He saw no reason at all to risk his life.

He and Monroe Payne opened the door to the store without a fight, watching the bandits select the supplies they wanted then

Battle of the Bloody Bend

helping them pack the merchandise up and carry it out to their horses. The Mexicans patiently and methodically took their time. They were safe in that faraway and isolated land. There would be no one to disturb or interrupt their raid, no one to take back what they had stolen.

One of the bandits wanted to kill Jesse Deemer.

Others refused. Too many times, the storekeeper had gone out of his way to help Mexican families when they were destitute and hungry. Deemer would live.

He would not be left at Boquillas.

The revolutionaries tied Deemer and Payne and carried them as hostages back across the border. They began a slow, tedious ride back toward the mountains that rose above a land where no one could find them.

Two days later, at Fort Clark, Colonel W. W. Sibley of the Fourteenth Cavalry received a telegram from Funston, directing him to assemble Troops F and H, as well as a machine gun troop, and depart as rapidly as possible for Marathon. The raids on Glenn Springs and Boquillas would not go unpunished.

In Marathon, Sibley and his men joined Major George T. Langhorne with A and B Troops of the Eighth Cavalry. Together, they would invade Mexico, even though the Mexican government was loudly claiming that the two offenses had been the barbarous works of lawless elements residing on the American side of the Rio Grande. Sibley later wrote that his orders were "to do my best without disabling men or horses, to run down and attack the bandits who had committed the Glenn Springs outrage . . ." He left his Cavalry unit at Bone Springs and traveled by automobile across the rough, weather-scarred roads that led to Boquillas, arriving to survey the damage and the carnage. A reporter would gaze across the Big Bend landscape and say, "The country isn't bad. It's just worse. Worse the moment you set foot from the train, and then, after that, just worser and worser."

At Boquillas, Sibley learned that the bandits had indeed retreated into Mexico and were moving slowly southward. They had a five-day head start. He and Langhorne had no time at all to waste. Besides, no one yet knew the fate of Jesse Deemer and Monroe Payne who were being held captive at Cerro Blanca. Langhorne was ready to move out. He took eighty hand-picked men, crossed the Rio Grande at the San Vicente ford, and began following a cold, worn-out trail. He would have benefitted greatly from the tracking talents of the old Seminole Negro scouts, but they had been disbanded two years earlier at Fort Clark, and Langhorne was on his own. Behind him, in strong support, came the Fourteenth Cavalry. Langhorne could strike and strike quickly. He wouldn't need strength, he decided, as much as he needed speed. The Cavalry would be his support only if a skirmish exploded into a full-scale war. Langhorne's motor truck rumbled and sputtered across the burning desert floor.

When Sibley arrived about six miles north of Tariases, he received a letter that had been scribbled by Jesse Deemer in El Pino. As the Colonel reported:

> He said he had been well treated and that the bandits were willing to exchange him and the Negro Payne for the members of their band captured by the mining men south of Boquillas.

A few of the bandits, on the morning of the raid, had wandered off long enough to rob the payroll from a silver mine near the base of the mountains. They piled the loot in the back of a truck, ordered three of the employees inside, and shoved the guards into the vehicle with them. Most of the revolutionaries rode on ahead. The driver, realizing that the Mexicans were probably ignorant of the truck's mechanical operation, told them that his motor had heated up. He eased along slowly as the old engine growled and whined beneath the hood. Then he shrugged and said, almost apologetically, that the vehicle was stuck in the sand. As the bandits pushed from behind the driver slammed the truck into reverse and knocked his captors to the ground.

Now they were captive in a dark Big Bend cell, and the raider leader wanted to trade Deemer and Payne for them. Sibley folded the message and stuck it into his pocket. At the moment, he was not in a lenient mood. He and Langhorne would first try to free Deemer and Payne on their own before they even considered negotiating for their release.

The colonel was much more worried over the note from Major Langhorne. He and his men were suffering from the heat and from lack of water. They had been slowed by bad roads. Sibley remembered:

> ... his plan to rescue Deemer could not be carried out as originally planned; ... men and animals were exhausted from hard riding, heat and dust, and were unable to make the necessary time.

On May 14, Langhorne's troopers would remain concealed at La Aguita during the daylight hours.

That night, they would attack.

Sibley had nothing he could do but wait. The day seemed a little hotter than before, and the hours dragged toward sundown.

Battle of the Bloody Bend

On the morning of May 15, the Colonel moved his Cavalry from Fort Clark to a small spring called Los Alimos, a strategical point that covered all trails over which the Mexican troops could enter the region. Shortly after noon, he received the message from Major Langhorne that he had been eagerly awaiting.

Sibley reported that the major had surrounded El Pino, early in the morning of May 15, but that the bandits had escaped, having departed the night before. Deemer and Payne, however, had been left behind by them under the nervous eye and gun of a Mexican rancher, "with the threat that unless the prisoners were kept safely until called for by them, he would pay the penalty with his life."

The bandits were not aware that the Cavalry was chasing them across Mexico, and they began burying themselves deeper into the high country, hiding away in canyons that lay upon the land like angry gashes. Sibley wrote that the mountainous region west of Sierra Majado was

> ...the center of activity of these bandits who owe allegiance to no party and who had but a short time prior to our advance ... attacked and defeated a command of Carrannista soldiers sent there as a protection to that locality.

This concerned Sibley.

His orders were simply to take action against those raiders who had wreaked havoc on Glenn Springs and Boquillas. And, as he reported, his men were not supposed to be "a police force to rid Mexico of its undesirables."

Already, the skirmishes had begun.

Sibley and Captain Goethe with his F Troop rode into El Pino about six o'clock in the afternoon, and a Mexican boy was waiting for him with another message from Langhorne. The Colonel would write,

> ... he (Langhorne) stated that one of his detachments under command of Lieut. Cramer, 8th Cavalry, on the evening of May 15th, had struck a band of twelve bandits at Santa Anita about seven miles west of Corro Bianca, had wounded five or six, capturing two of the wounded and had brought in about 15 ponies and small mules with considerable of the loot taken from Deemer's store, together with ten stands of arms and ammunition.

He joined Langhorne, and the two Cavalry officers rode twenty miles further south as their men searched the trails and the foothills of the mountains. But all traces of the raiders had faded, had been blown away by the Mexican winds.

In his report, Colonel Sibley would explain:

From . . . other information received after my arrival at El Pino, I became convinced that the bandits who now all knew of our presence in the country, had been driven so far south and had been so badly scattered and punished that our mission as outlined in my instructions . . . was completed . . . Attention is invited to the actual loss of these bandits since the beginning of their attack at Glenn Springs. An authentic report from Mr. Deemer places the size of the original band at fifty-one. At Glenn Springs there is every reason to believe that four or five were killed, the bodies of two having been found, and a number wounded. At or near Boquillas, Mexico, three were captured and are now in jail at Alpine. Four or five were wounded and two of the wounded captured at Santa Anita by Lieut. Cramer and a detachment of the 8th Cavalry. All reports show that the band was completely broken up, only a few being reported as still together and these were fleeing towards Sierra Majado. Another of the band who admitted that he had been present at Glenn Springs, wandered into camp of one of the detachments near Cerro Blanca, and gave himself up, saying that he had been without food or water for three days. Of the original fifty-one who made this invasion at least twenty-five per cent had been killed, wounded or captured, and they had lost in addition ten rifles together with a lot of their loot and a number of mules and ponies.

Great difficulty was experienced in supplying this command, badly scattered as it had to be on account of scarcity of water due to the breaking down of the wagons and automobiles, all of which including Maj. Langhorne's private car and one government auto truck had to be left on the road in the advance to be picked up on our return . . .

The officers and men . . . are entitled to the highest commendation for the thorough manner in which they performed this arduous duty, and for the cheerfulness with which they bore great discomforts and suffering from heat, lack of water, short rations and long marches, and yet returned with no casualties, no sick, and no sore backs among the horses.

9

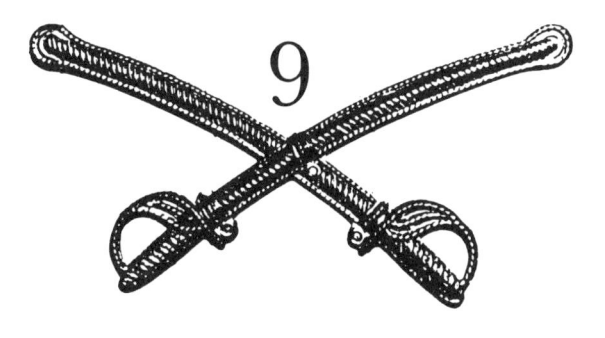

The Coming of Age

America had been at war in Europe, facing the Kaiser on battlefields far removed from the North American continent. The Cavalry of Fort Clark, as usual, was more concerned with Mexico. But then, that's the way it had always been. Peace had never come to stay for very long upon the muddy Rio Grande.

In 1918, as the armistice was signed to end the bloodshed of World War I, work got underway to greatly upgrade the facilities at Fort Clark. All talk of abandoning the outpost had long been silenced, drowned out by the fighting in Europe and the gunfire at Glenn Springs. Fort Clark, it appeared, did indeed have a role to play in the United States military as it stood firm against aggression around the world. The post was enlarged with many of its aging, decrepit buildings —survivors of the Indian wars— undergoing much needed repair and renovation. And a year later, Major General John W. Ruckman dedicated the Y.M.C.A. at Fort Clark, described as one of the army's best, housing it in a structure that one day would hold the Service Club.

By the time Major H. H. Smith of the Medical Corps made his report in 1927, Fort Clark had grown to include fourteen sets of officers quarters, six N.C.O. quarters, ten two-story barracks, five administration buildings (including the hospital), ten stables and five storehouses constructed of a soft stone from a quarry on the post itself. The major explained that:

> Officers quarters are of stone, the buildings being some eighty years old, having old fashioned fireplaces which are aided by wood heating stoves. All that can be said for these quarters is that they are serviceable and sanitary and that they meet the conditions, and to recommend any improvement would be to rebuild the post from the ground up.

In addition, soldiers had the benefit of a barbershop, post exchange, commissary, laundry, theater, restaurants, and a swimming hole.

Fort Clark was slowly coming of age.

However, it still had its share of problems.

The clothing, Major Smith said, presumably issued from war stock, was "shoddy," "ill-fitting and coarse in texture." The cotton khaki, he pointed out, was "of all colors and the colors of some changes with each week of laundry." The wooden pipes, carrying sewage away from the post, had rotted and begun to leak. The dental officer had been sick and off duty for some time, so those with toothaches just ached and bit the bullet, and kept right on with their mounted drill. An epidemic of scarlet fever had swept Fort Clark. But Major Smith was much more concerned about the high rate of venereal disease among the troops. He reported,

> The entire command spent the greater part of the summer at San Antonio where the soldiers were exposed to unrestricted venereal conditions not within the province of the Medical Department, also Mexico is thirty miles away and under no sanitary regulations.

His solution to the problem was a simple one: merely establish a prophylactic station at the bridge heads of both Eagle Pass and Del Rio. The men might not be able to control their lust, but the army could certainly try to control their amorous diseases.

Life at Fort Clark had become routine, even dull. Only the venom of rattlesnakes remained as a threat upon prairies that no longer defied the stumbling advance of civilization. Now automobiles even raced where horses once feared to tread. And boredom was the one enemy that couldn't be whipped nor ignored. Soldiers cleared about fifty acres along Las Moras Creek, irrigated them and began raising fresh vegetables. Others sold their garbage to people in Brackettville for hog feed. The town itself still only had about 500 residents and, according to one army report, ninety percent of them were "dependent on Fort Clark for their livelihood." This village, it pointed out, "contributes practically nothing toward recreation or uplifting influence for the troops at Fort Clark."

Commanding General LeRoy Elting worked hard at Fort Clark to maintain a neat, orderly appearance of the post — particularly the officer's line. Lawns had been seeded and rose gardens planted, and a new cement sidewalk had been paved along the row of houses. The general was extremely proud of his Post Exchange, which included a lunchroom, tailor shop, barbershop, shoe repair shop, and a market for fresh meat and fruits. However, Elting also knew that a severe shortage of money had drastically reduced his ability to make life at Fort Clark really bearable for his command: fifty-seven officers, three warrant officers, thirty-two non-commissioned officers, 1,724 enlisted men, and 2,043 animals.

As Lieutenant Colonel George L. Hicks, the assistant adjutant general, had reported:

> The lack of funds for repair and administration has thrown an excessive burden on troops, has seriously interfered with training, and has been the source of much justifiable complaint. It has been necessary to limit training and care of animals to three hours a day. An average of about five hours a day for each enlisted man has been spent on fatigue, construction and repair, made necessary by lack of funds to hire civilians for these purposes. While eight hours a day can not be considered an excessive amount of work, it is believed that the average soldier enlists to be a soldier and not a laborer. He does not object to work in connection with his training as a soldier, but does resent work entirely disconnected with training, and which was not mentioned in the recruiting advertisement which induced him to enlist.

There was also criticism from Colonel Hicks because the absence of officers and enlisted men who attended horseshows and polo tournaments had likewise seriously interfered with their training and performance of duty.

The men had become pawns — used, misused, and sometimes abused. Few learned to fight at all. After all, the post's 3,963 acres were much too small for any combined maneuvers of simulated warfare. The airplane landing field had been dug out from among the chaparral and mesquite groves. The target range had no backstop, which made firing unsafe on the reservation. A stray bullet during training could be deadly, and most of the bullets went astray. So the soldiers spent their hours trying to nail rotting boards back together to keep out the winds of winter, and trying to patch the leaks in their paper roofing.

Fort Clark had become an orphan stepchild of the army. It had two battalions of men, including the Fifth Cavalry. Most of the troops, during this languid time of peace, were merely putting in their time, tucked back in a forgotten corner of the world that was crumbling all around them. It was as though they had been exiled for the duration of their enlistment. No one in power, it seemed, remembered them or cared about them and their comforts at all.

Again, the military began fanning the flames of old rumors that Fort Clark would soon be abandoned. It had become an expensive blight that the army wanted to delete from its records and its responsibility. Only the bandit raids from Mexico had saved the post from extinction back in 1912, when the fort had existed as a virtual ghost town. For a time, it had been needed. Now it wasn't, not anymore. Now it appeared to be doomed, as military documents kept censuring an outpost that age had made obsolete.

Some had grumbled that the brightness of the yellow stone, glistening in the sun, was hard on their eyes. So an effort was made to ameliorate the condition by "blue washing" the buildings. Temporary barracks were thrown together to house a machine gun squadron, but the army said that they were unsafe, dangerous, and should be destroyed or salvaged. The stone buildings were decaying and had to be bolted together by iron rods to keep the walls from falling out. The mortar and plastering had lost their strength, and there was a constant sifting of soft dust and sand from the deterioration. Many of the old barracks were lined with cracks, and one officer lamented that the buildings were obviously "put together by soldier labor or unskilled native mechanics." The woodwork had become so full of dry rot that two nails could not be driven with security into any of the splintered beams.

Every faucet at Fort Clark was either dripping or leaking, and half of the toilet bowls in the latrine were out of order. The electric wiring was unsafe, and in the stables, the wiring had become positively dangerous. Asbestos shingles salvaged from Camp Eagle Pass were being shipped to Fort Clark to replace the wooden shingles that had rotted away from the roofs. Of the seventeen frame barracks, sixteen two-story buildings were in such poor condition that they were no longer considered safe during a storm. "They should be razed as soon as vacated," the report said, "and they should be vacated now."

It also pointed out, "A careful examination of a couple of noncommissioned officers quarters being erected at Fort Clark from salvage material from Camp Eagle Pass showed that it is a combination of the patchwork worthy of habitation by the lowest type of day

The Coming of Age

laborer." In the Inspector General's opinion, the shacks should have been rented or sold and left on the border. It was a disgrace for a military man to be subjected to such low-class shelter.

Fort Clark had three five-passenger Dodge trucks, a dozen cargo trucks, and one fire truck. A mechanical inspection indicated that all of the vehicles were in need of repair, some being in a salvage condition, and others requiring a complete overhaul. Blame was placed on "untrained and inadequate personnel," as well as on the "lack of proper storage facilities." The fort depended on those trucks and could not really afford to have them broken down under some shade tree. Those vehicles, along with wagons pulled by draft mules, were used to haul supplies to Fort Clark from the Southern Pacific Railroad yard at Spofford Junction, located about ten miles away. They limped along over a graded road that was pockmarked with chuckholes, a pathway that was "very dusty when dry and almost impassable to motor transportation when wet."

And the incinerator hadn't been used at the post for more than a decade. It wouldn't work for the lack of draft. So the garbage at Fort Clark — that not sold for hog feed —was carried to a dump on the western edge of the reservation and burned.

Major E. P. Pierson, the Inspector General, reported:

> There is no doubt that in the frontier days it was a strategic move to establish a post at the source of such an abundant supply of water as the Las Moras Spring represents. The surrounding country is extremely dry and artificial water holes were unknown. There are similar sources of water, the nearest important one being at Del Rio, 30 miles west, while the Nueces River, about thirty miles to the east, furnished a stopping place for marching troops. These were all matters of importance in the days preceding the Southern Pacific Railroad. Its location never had any tactical value. In fact when the railroad chose a right of way leaving Las Moras Spring 10 miles away, this spring ceased to have any but a local value, since the railroad yards and roundhouses were constructed at Del Rio. Until recent years, when the land became relatively well settled, there was some necessity for troops to be concentrated at Fort Clark as a threat against marauding Indians and Mexicans. These conditions have long since ceased to exist and Fort Clark is one of the Posts in this Corps Area that has the distinction of having been abandoned and neglected until almost uninhabitable and then to be continued to be occupied from necessity to find a place of comparative shelter for the troops. The world war coming on required the use of every available facility controlled by the War Department. It was therefore natural that some temporary structures were added at Fort Clark and that some of

the modern facilities such as sewer and electric lights were installed also in the old buildings. However, the wartime temporary standard of construction was used and all of these now need reconstruction to insure safety and sanitation.

Therefore the value of Fort Clark is largely only its land value which is estimated at $60,000. Very few of the buildings have any value at all except for occupancy by people of very much lower standards of living than those desired in the army.

Major Pierson had come to one conclusion. Fort Clark should be abandoned and sold, with the military post being moved to Eagle Pass. Every dollar spent repairing old buildings at the post, he said, was being wasted. The army would be wise to start anew somewhere else. After all, he reasoned, the little town of Eagle Pass already had such utilities as water, gas, and electricity. It was situated much more strategically on the border of Mexico. And its people were friendly toward troops, were even actively lobbying to have a fort brought within their midsts.

Brackettville was symbiotic with Fort Clark. Yet, while depending on Fort Clark for survival, it rarely gave it anything in return. General Elting even indicated that the Southern Pacific Railroad owned 40,000 acres near Eagle Pass that would be "of suitable character and location for a post." He suggested,

> It is quite possible that the Southern Pacific Railroad would be willing to give the surface rights to 5,000 or 6,000 acres of this land and lay a spur track to it from the railroad station at Eagle Pass in exchange for the Fort Clark Reservation, with a view to moving their whole plant from Spofford Junction to Fort Clark . . . the Fort Clark reservation though worth only $40,000 to $60,000 at a forced sale should be worth as much as $150,000 to the Southern Pacific . . . while the bulk of their holdings near Eagle Pass is not worth over $10 per acre . . .

The fate of Fort Clark wavered.

It was a shame, an unwanted feeling that the old outpost had experienced so many times before.

In a 1906 editorial, for example, the *Brackett Mail* said:

> *The Daily Express*, in conjunction with Congressman Slayden and the merchants of San Antonio, has been working for the enlargement of Fort Sam Houston, and to attain this end is fighting and has fought Brackett for years in every attempt she made towards the retention of troops or the upbuilding of Fort Clark. The *Express* says 'that the removal of the 1st Cavalry is prophetic of the abandonment of Ft. Clark.' The *Express* has posed as a prophet for

The Coming of Age

> long years, and its prophecies have always portended evil to this western section of the state. San Antonio derives an immense trade from this western country that she is going to lose. The merchants are getting tired of paying her tribute and having their throats cut at will. It is time for us to take up the fight and let it be to a finish. Let the watchword be 'On to Houston!'

Fort Clark appeared to be doomed.

And the newspaper tried to brace the citizens for life without the outpost, trying to infect the town with a spirit of hope and pride. One article pointed out:

> If Fort Clark is abandoned Brackett can still fill all vacant houses with a desirable class of families by building a new school house and keeping up the high standard of the school. With a new building and with about $75 or $100 spent in judicious advertising it will not be difficult to induce many new families to locate here. Sonora, Ozona and many other towns flourish without a Post and why not Brackett?

Within the administration of President Theodore Roosevelt, efforts were definitely being made to shut down Fort Clark as a military post. As former Vice President John Nance Garner of Uvalde recalled, "He (Roosevelt) and old man Taft, then Secretary of War, were planning to pull the soldiers out of Fort Clark and [then] guard the border from Santone."

> I flagged my coattails up there in my best rookie congressman manner and told them it would be a sin to waste the best hosses, the best likker and the prettiest wimmin of any place on the border.
> They both laughed like hell — and revoked the order.

Fort Clark had again bitten the bullet.

By 1927, Brackettville had grown to a thriving population of 1,600 souls who sometimes occupied the Catholic, Methodist, Episcopal, and Christian churches that had been wedged between the saloons and houses of ill repute. There were places to sin, and plenty of preachers to save the sinners.

Cattle still grazed the sprawling ranchlands, and the economy depended chiefly on the sales of livestock, wool and mohair, pecans, and hay. Some sheared sheep for a living, others picked cotton, and many gathered honey. Chicken farms and dairy farms dotted Kinney County. Almost every little house had a garden out back. Brackettville's volume of business began flirting with the million dollar mark.

A little later, prosperity was wearing thin.

In 1929, the stock market crashed, but Fort Clark, the old outpost on Las Moras, didn't crash with it. The Army again stuck with the isolated sentinel on the prairie, even though the economic struggles of the Great Depression crept across the countryside like a runaway epidemic that could neither be stopped nor controlled. But then, Fort Clark had faced troubled times before. In fact, it seemed to thrive on crisis.

When Lieutenant Colonel G. Kent inspected the outpost in 1930, he was able to report:

> The general police of the Post was good, the maintenance and repair of buildings and utilities shows intelligent and careful supervision by commanding officers and quartermaster. The appearance of the officers, enlisted men and animals was excellent and the morale of the garrison was good. From conversation with both enlisted men and officers the inspector is of the opinion that the state of mind of the command was one of contentment.

To him, the Quartermaster Department appeared "to be operating efficiently and economically." The warehouses and storerooms were found to be satisfactory. The barracks and stables were well policed and cared for. Graduates from Cooks and Bakers School were even being utilized in the kitchen. "Moving pictures, dances, and sports," he wrote, "are encouraged and lend to the contentment of the command." The Colonel also pointed out that "the Post Exchange is a great convenience to this Post, situated as it is, with no good stores in the vicinity." He did, however, complain that the heating plant in the hospital was antiquated and inadequate, that the quality of beef received for the tables was very low, and that general prisoners were required to stand at attention with folded arms which was a definite violation of the Army Regulations.

Colonel Kent was also concerned because two windows in the guardhouse had lost their glass panes, permitting a cold wind to blow on the prisoners while they took their baths. It also made him remember the item which had appeared long before in the post newspaper. Company D had lost the Battle of the Bath against Post orders commanding that all soldiers should bathe every morning. And it was reported: "Killed — none; wounded — 'Texas pride'; lost — estimation of Post Commander."

Each morning, a special detail would march into the guardhouse with a hose and wash the prisoners down. The jailbirds got all of the water they wanted, usually from the nozzle of the hose, and a loaf of bread. And, of course, they hid their chewing tobacco in those knotholes that marked the wooden floors beneath their bunks.

The Coming of Age

Fort Clark, during the 1930s, regained a certain measure of esteem and respect that it had lost after its glorious escapades during the Indian wars. There was no infantry at all at the post, only horse soldiers and 2,600 mounts. The barracks had finally been condemned and, until 1933, the Cavalrymen lived in a tent city, and none of them seemed to mind. They didn't have time to complain, and generally they were simply too tired to bitch about their misfortunes.

The troops were up at 5:30 in the morning, and the whistle blew to fall outside promptly at twenty minutes to six. They ate quickly, marched briskly, then gathered up their gear and reached the stables by eight o'clock. Horses were tied to the picket line, groomed and brushed and saddled up. On the way to rifle range, one recruit remembered, "We walked 'em and trotted 'em and kept 'em physically fit like a well-trained fighter." At the end of their mounted drill, the men all had to clean and saddle soap their own equipment.

Each officer selected a man, called a dog robber, to be his personal orderly. It was an honor. It was work. That trooper's feet hit the ground at five o'clock each morning. He would hurry out across the parade ground to the officer's quarters, build a blaze in the fireplace, cook breakfast, and make sure that all boots, belt, and guns were polished and in place. At the stables, the orderly called up the officer's horse —always a thoroughbred — groomed him, put wax on a cloth and buffed the mount until it glistened in the hot Texas sun.

There was pride in the Cavalry.

Yet the need for horse soldiers was slowly dying away from the military's board room of strategic planning.

Charging into battle on horseback was becoming obsolete, whether Cavalrymen wanted to believe it or not.

The horse soldiers had two basic objectives at Fort Clark. They learned to ride and ride well. Every man was trained with an air-cooled machine gun. They knew how to fire it, take it apart, clean it, and put it back together again. As the drill sergeant kept telling them, "Know it as well as you know your own family. Someday, if war comes, it can save your life." The men trained and sweated and cursed the harsh earth and the harsh sun, but they didn't worry much about going into battle. Why should they? After all, the war to end all wars had already ended.

They fiddled with chopsticks at the post's Chinese restaurant, bought hot bread from the bakery for three cents a loaf. With a ten dollar bill they could buy more groceries at the commissary than they could carry out. Gasoline cost them nine cents a gallon. And old Mr. Martin, a Kinney County farmer, always hooked up his team of mules and had them ready when the rains began to fall across the brush country. Automobiles, straining and grinding, just couldn't make it through that black gumbo prairieland. Mr. Martin charged the soldiers a dollar to pull them out of the mud when their cars got stuck, and they almost always did. Their thoughts were more evil than their words, because one officer always told them as soon as they set foot on Fort Clark: "Put on white gloves and get all the nasty words off your tongue. There'll be no more cursing." He meant it. From then on, the men muttered their various and sundry blasphemies under their breath.

For most new recruits, Fort Clark was an experience that they didn't quite expect.

George Wyrick had been a high school dropout from a little cotton mill town in North Carolina, when he decided to cast his lot with the army, and most of his life had been spent on a farm, usually behind a team of stubborn, ornery mules. He was standing in formation in Fort Bragg, feeling lost and out of place, when a sergeant barked, "Any of you young men who have ever been farmers or handled animals and want to go to Texas, take one step forward."

Wyrick recalled, "I almost broke both legs getting out to where he could see me."

A few days later, he was sitting on a train headed west, arriving at Spofford Junction during a warm summer rain. Wyrick was marched outside and ordered to load his gear onto the back of a truck that had solid rubber tires.

His gear was dry when it reached Fort Clark.

He wasn't.

"We had to push the damn truck all the way to the fort," he said. Those rubber tires had been sucked up by the mud and the black gumbo that kept old Mr. Martin in spending money.

George Wyrick's first impression of Fort Clark was probably typical of most recruits who suddenly found themselves stranded down along the Rio Grande. He said,

> It was desolate. All I saw was a big open parade ground, and it looked like a desert. And when we marched, all our feet ever touched was rocks. I thought, 'Oh, hell, it could rain for forty days and nights and I'd never even get my feet muddy.'

The men had it easier than the automobiles when the rains came down to wash the dust away from the face of Fort Clark.

The men found their entertainment at the Post Theater, or they would venture seven miles down to the dance hall at Cline on Sat-

The Coming of Age

urday nights. They swam in the swimming hole beside Las Moras Springs, out amongst the weeds and stones where the snakes had a habit of sunning themselves. Once a month, they would walk to the commissary and climb three flights of stairs to pick up their $16.25 from the finance center. The pay, as always, was disappointing. But a sergeant told George Wyrick, "If you go and become a pilot, you can draw more money."

"So," he said, "For a few years I became a *pilot*. We'd go down to the stables, and we'd *pile it* here, then *pile it* over there."

When Prohibition ended, so did the doldrums.

Brackettville, adept at understanding the free enterprise law of supply and demand, expanded its downtown to include sixty saloons and four houses of ill repute. And that $16.25 each month didn't last as long as it had when the men had sneaked illegal whiskey from the back pockets of backdoor peddlers. As Wyrick recalled, "If things got dull, somebody would work up a good fight." There were a lot of fights. Fights were free.

Down on Mulligan's bend of Las Moras Creek, the girls from San Antonio, with cheap perfume and faded smiles, would gather with their well-worn bedrolls each payday to offer comfort and solace to soldiers who weren't in the mood for dancing and who had lost their thirst for whiskey. The men were a long way from home. They were lonely. The night hid them. All they needed was a little love, and love only cost a dollar — or sometimes two. And the perfume wore off about as quickly as the memories. Within three days, the money would be gone, and so would the girls.

A sign on the roadway that led into Fort Clark said, WHEN YOU CROSS LAS MORAS, YOU HAVE YOUR SINS WASHED AWAY.

But, the soldiers reasoned, *it didn't say you couldn't pick them up again when you headed back across the creek.* They went back across as often as they could.

If a trooper had too much to drink and his head hurt much too badly for him to fall out for roll call, an understanding sergeant would merely walk into the barracks, pick the aching soldier up in his arms and throw him outside. He usually spent his next Sunday on KP duty or with the stable clean-up crew. The habitual complainers were always drawing extra duty. The First Sergeant studied each man until he knew him like a book. As Wyrick remembered, "If a guy soldiered like hell all month, got paid, stayed drunk for a couple of days till his money was all gone, the officers didn't get upset or worried at all." They understood. There was no reason to get angry. Sometimes all they needed was to blow off a little steam and rub their problems and their loneliness away with a little alcohol.

But they did get concerned when Pop Nelson turned up missing. Pop Nelson was the mess sergeant, and a good one, and that made him the most valuable person at the fort. For the morale of a unit sometimes depended solely on the grub they ate. If their empty bellies were satisfied, they could usually be trusted to do a good day's work.

When General Jonathan Wainwright came to Fort Clark, he brought a little black bear cub with him, even built the little critter its own personal den. It was Pop Nelson's duty to take care of the cub. And he fed the little bear and watched him grow.

One day a payday came.

And Mess Sergeant Nelson vanished.

The next morning, food was missing from the kitchen. For five days, troopers searched in vain for Pop Nelson. For five mornings, food had been stolen from the kitchen shelves.

On the sixth evening, a soldier walked past the black bear's den and saw a lifeless human foot sticking out from the shadows. Fear gripped him, then nausea. Pop Nelson, the soldier imagined, had fed the bear once too often, had gotten too close to him for the last time. He ran to find the First Sergeant.

The soldiers kept the bear at bay while a detail sought to gingerly remove the ruins of Pop Nelson.

He wasn't ruined.

He was snoring.

And there beside him was a gallon jar of whiskey whose contents had kept the mess sergeant unconscious and gloriously at peace with the world for almost a week. Pop Nelson had needed to get away for awhile, and he had found the perfect hiding place. The bear had neither bothered nor harmed him. The bear had taken care of him. As far as the young bear was concerned, Pop Nelson was his mother.

General Wainwright had a steadying influence on the men of Fort Clark. He was, they believed, an enlisted man's general, and they loved him. Wainwright was fair, and they knew they could depend on him. The general simply saw that the soldiers at Fort Clark were fed, clothed, and taught how to fight. George Wyrick recalls, "General Wainwright always gave you the best he could under the circumstances. If you wanted a fistfight real quick, just say something bad about him."

No one did, at least not loudly.

Late one cold, rainy night, General Wainwright and his chauffeur were driving around the post, informally checking the mounted guard. As they slowly approached the Quartermaster area, a young man rode out from the darkness with his pistol raised, and he shouted, "Who goes there?"

The chauffeur promptly stuck his head out of the car window, and wiping the rain from his face, he yelled, "The *General!*"

The guard didn't waver.

"Will the general dismount, advance, and be recognized," the young trooper demanded.

"It's *raining*," the chauffeur shouted.

"Then let the general get *wet* and be recognized," the guard snapped.

With a sigh, General Jonathan Wainwright unfolded himself and crawled out into the chilly downpour, trudging through the mud toward the guard.

He looked up.

The young cavalryman paled. He was staring down upon too many stars upon the officer's shoulders. He lowered his pistol and nodded. He couldn't speak. His voice, as well as his wits, had left him.

The next morning, General Wainwright immediately called the young man's sergeant to his office and made sure that a three-day pass was waiting for the pale soldier when his guard duty had expired.

The general hadn't minded the rain, the cold, or the misery. He was, however, proud of a young man who had admirably performed his duty regardless of the consequences.

Wainwright's men believed in him, and they protected him every chance they got.

During competition to decide the best platoon of Cavalry at Fort Clark, a group of officers rode proudly out onto the parade ground to be introduced. Wainwright led the way. He was to be the judge. The ultimate decision belonged to him. But the effects of a happy-go-lucky drinking bout the night before had not worn off, and the world kept spinning, and his horse, it seemed, kept right on running even when he was standing still. As Wainwright rose regally in his stirrups to speak, he slowly slid off his horse. An orderly frantically ran to his aid, grabbed the general, and helped him back onto his mount again. Wainwright sighed, grinned, and promptly fell off the other side of the saddle.

Only for a moment was there an awkward silence.

Discomfort had pervaded the parade ground.

Men stammered and stuttered and scuffed the ground with the

toes of their boots. Then suddenly, as the orderly lifted the general to his feet, a dashing, handsome young officer rode beside Wainwright, shielding his calamity from the view of the crowd, and he announced in a loud, firm voice, "I'm Major Gray. General Wainwright has just appointed me head umpire."

The competition began.

Wainwright walked proud at Fort Clark. He walked tall. And none of his fellow officers or men mentioned the incident to him at all.

During the 1930s, Colonel George S. Patton received his orders, assigning him to the Cavalry post at Fort Clark. For Patton, the duty would not be unpleasant at all. He was a soldier, and hardships were merely a way of life. They would merely harden him, toughen him for the years that lay ahead.

When he arrived on Las Moras Creek in 1938, a mounted review awaited Patton. He stepped boldly forward, looked around at the men who stared emotionless at him, and said, "I am most happy to be assigned here at Fort Clark with the Fifth Cavalry. I understand that you soldiers are the world's greatest horsemen and greatest fighters, and — above all —damn good lovers."

The men smiled.

They relaxed.

They were facing a leader, and they would follow him anywhere, anytime. George Wyrick would recall, "Patton wasn't the kind of man who would say, 'Lieutenant, you take your outfit and go do this job.' He'd say, 'Lieutenant, go get your outfit and come with me.' "

He would fight for them.

Early one morning, the troops left Fort Clark, marching toward Camp Bullis, and Patton had taken an advance party out with him to control the flow of traffic on the highway that ran past the post.

The men were moving along both sides of a dirt road, and a car came barreling down between them at a reckless speed. Patton, almost without thinking, turned his horse out in front of the speeding automobile. The mount was prancing nervously, but Patton stared straight ahead with a stone face, his jawbones gripped tightly together in anger.

"Stop that car!" he shouted.

The driver braked to a fearful stop, sliding on the dust and rocks.

"Young man," Patton barked, "This is the Fifth Cavalry of the United States Army. You park that damn car and don't move it until the last horse is past you."

The Coming of Age

The driver nodded.

Patton rode away, and the automobile didn't budge at all until all 2,000 horsemen had trotted beside him and were out of sight. He didn't see Patton again. He didn't want to. Patton was tough. And *no one* doubted it.

There was an uneasiness in Europe that bothered Patton when he assumed the reins of command of the Fifth Cavalry. Germany had already begun testing its new weapons and tactics on the plains of Spain. Hitler was raving like a madman whose thirst for power could very well launch a full-scale war upon the smaller nations that hid with trepidation just beyond Germany's borders.

The United States ignored the rumblings, the faraway hint of war.

George S. Patton didn't.

He simply awoke Fort Clark from its lethargy. Units were sent out in the prairies to practice miniature war games, working day and night to develop new techniques of a dismounted attack. Patton was a cavalryman, the last of a breed. And he could foresee that horses and mules would never play a major role in warfare again. His men, if they were to survive the front lines of battle, would become machine gun companies.

From the stables of Fort Clark, the troops rode horseback out to their designated battlefields among the mesquite and bramble bush of the Rio Grande.

But they fought on foot.

And they grumbled because George S. Patton, in his arrogance, even demanded "spit and polish" on the wind-blistered sands of a wind-blistered desert. He heard them, but paid no attention to their complaints. He was a disciplinarian. The military habits that became a way of life at Fort Clark would someday save their lives on the foreign fields of Europe.

The men might curse him.

But, by damn, they would respect him.

Patton drove his cavalrymen, almost without mercy. Their skills sharpened, and so did their dedication. They were raw recruits no longer. They had grit, and they became a defiant fighting machine.

Patton was loved.

He was hated.

He was domineering.

He knew how to win, and in war there was no alternative to him. As he one day would tell his troops: "No bastard ever won a war by dying for his country. He won it by making the other poor dumb bastard die for his country."

Perhaps Patton was hard on his men at Fort Clark. He had a reason. Patton simply wanted to see how each of them reacted under pressure — just or unjust. Later he said, "I get most of my best men that way."

He strongly believed:

> The soldier is the Army. No army is better than its soldiers. The soldier is also a citizen. In fact, the highest obligation and privilege of citizenship is that of bearing arms for one's country ... Anyone, in any way of life, who is content with mediocrity is untrue to himself and to American tradition. To be a good soldier a man must have discipline, self-respect, pride in his unit and in his country, a high sense of duty and obligation to his comrades and to his superiors, and self-confidence born of demonstrated ability ...
>
> No sane man is unafraid in battle, but discipline produces in him a form of vicarious courage which, with his manhood, makes for victory. Self-respect grows directly from discipline. The Army saying, 'Who saw a dirty soldier with a medal?' is largely true. Pride, in turn, stems from self-respect and from the knowledge that the soldier is an American ...

Patton never pushed his men.

He led them.

That, he was convinced, was his most important responsibility of all. He later wrote his wife, shortly after an invasion of North Africa in the fall of 1942:

> On the morning of November 9, I went to the beach at Fedhala accompanied by Lieutenant Stiller, my aide. The situation we found was very bad. Boats were coming in and not being pushed off after unloading. There was shellfire, and French aviators were strafing the beach. Although they missed it by a considerable distance whenever they strafed, our men would take cover and delay unloading operations, and particularly the unloading of ammunition, which was vitally necessary as we were fighting a major engagement, not more than fifteen hundred yards to the south.
>
> By remaining on the beach and personally helping to push off boats and by not taking shelter when enemy planes flew over, I believe I had considerable influence in quieting the nerves of the troops and on making the initial landing a success. I stayed on the beach for nearly eighteen hours and was wet all over all of that time. People say that army commanders should not indulge in such practices. My theory is that an army commander does what

The Coming of Age

is necessary to accomplish his mission, and that nearly eighty percent of his mission is to arouse morale in his men.

Patton had found himself in the awkward position of fighting French troops when he stepped ashore on North African soil. It troubled him. After all, for decades, France and the United States had always been friends in time of peace, and allies when war raged around them.

When he was at last handed a set of surrender terms at Fedhala, Patton looked sternly at the French commander and said, "We are both officers and gentlemen. I respect your valor and that of your men. We have worked and fought together in an earlier war. If you will give me your hand and your word that there will be no trouble between your people and mine, this will be surrender terms enough."

Patton, with a theatrical flair for drama, staged a full dress military funeral on the embattled beach. An American soldier and a French soldier were symbolically buried in a common grave, their coffins draped by the joined flags of both nations.

Patton did indeed know how to arouse the morale of his men. It was a lesson that he had learned early in his military career. Being a leader of men was also a lesson Patton had learned early in his military career.

It was a lesson he practiced daily on those field maneuvers at Fort Clark.

George S. Patton was no stranger to the desert country. He had journeyed there with John J. "Blackjack" Pershing, chasing the bandits of Pancho Villa far into the rock-scarred interior of Mexico. Bullets met them on the outskirts of one small adobe village, and Patton's touring car swerved off the dirty roadway as the young officer leaped to the ground, firing both ivory-handled pistols as he raced toward a crumbling mud-plastered wall. One of Villa's lieutenant's fell. Another, Cardenas, would neither surrender nor run. Gunfire cracked through the streets as chickens scurried for safety and mangy dogs howled and barked. At last, Cardenas was silenced, the ashen dust seeping up the blood from his wounds. Patton stood above him and said reverently, "The man had nerve."

He strapped the bodies of both bandits onto the fenders of his old touring car, intent on returning them to Pershing's tent. The soldiers eased through the wasteland, much like deer hunters coming home with their prey, and the sun beat down relentlessly on two of them. As Patton would later remark, "In that temperature, it wouldn't have taken long for a couple of Mexican stiffs to drive a buzzard off a gut-wagon."

He had no choice.

The men pulled the car over to the side of the road and began digging a couple of graves in the hard rock. Patton had one serious concern. There was no one among them who knew a funeral service to say over the bodies of the deceased. Prayer did seem like the Christian thing to do at the time.

Finally, a grizzled old sergeant looked toward Patton and said solemnly, "Don't worry, Sir. I know what to do."

Patton stepped back and bowed his head.

The sergeant stood looking down upon the mounds of dirt, and he recited:

> Ashes to ashes, and dust to dust,
> If Villa won't bury you,
> Uncle Sam must.

Minutes later, the touring car was gone, and the dust had begun to settle once again upon a land too alien for anyone to dwell on for very long. Patton looked proudly at his twin pistols. They had not failed him. He told a friend who asked about his pearl-handled revolvers: "Dammit, my guns are ivory-handled. Nobody but a pimp from a cheap New Orleans whorehouse would carry one with pearl grips."

Patton had always been a crack shot, one of the army's best marksmen. Shortly after his duel with the bandits of San Miguelito, he wrote his wife, Bea, and said: "You are probably wondering if my conscience hurts me for killing a man. It does not. I feel about it just as I did when I got my swordfish: surprised at my luck."

As a cavalryman, George S. Patton found spare time at Fort Clark to indulge himself in hard-riding, sometimes tumultuous games of polo. Anytime a team came down from San Antonio, he was ready to saddle up. For Patton, polo was almost a masochistic sport, with man and beast throwing themselves into the midst of turmoil and collision at high speeds, the ecstasy and agony of conflict. He asked for no quarter. He gave none. As George Wyrick remembered, "Patton wasn't just a rider. He was a player. He didn't mind bumping you in the heat of action. He didn't mind knocking you and your horse both halfway off the field." That's the only way he knew how to play.

Patton would simply smile and say:

"I love polo. It reminds me of war."

So did Europe.

The heavy, ominous sounds of Nazi boots were growing louder, echoing all the way to Fort Clark. America was suddenly having trouble ignoring those warnings of war.

The Coming of Age

The trick maneuvers — with the Twelfth Cavalry guarding the Rio Grande against illegal crossings of Mexican cattle, goats, and sheep, just didn't seem so important anymore. The military maneuvers did. Patton made them as realistic as possible.

During the Third Army war games, testing troop mobility, concentration, and deployment, he wrote his wife:

> This is the second day of the war. Yesterday we marched 35 miles in the worst heat I have ever seen and secured our objective without a fight. Said objective is a stunted oak forest on a ridge but owing to ground rules we have to sit in a windless valley. I have been here ever since 5 A.M. I think the infantry won't get up for another 18 hours and as the enemy who is 4 miles north of us won't fight I guess we will just sit. Tomorrow night we may march around his flank but it will be a very long trip over slippery roads. Still at night it is cool.

A few days later, Patton was feeling much better about his role in the simulated warfare. He wrote:

> We are now at Sabinal which is sixty-two or three miles from Clark. So as I am making two thirty-mile marches we will be in on the 25th [of August, 1938]. Actually, I am at Clark now having driven in the regimental car to get a bath which God knows I need...
>
> The last day of the Maneuvers I had a swell time. The 12(th) was in front on a flank march which I had advised the first day and was only put over on the fourth. It got held up for three hours and then they put B (Troop) and the Machine Gun troops of the fifth in the advance guard. I went along as sort of huntsman. We moved about ten miles mostly at a gallop and one horse died but we got right into the enemy rear areas and captured two battery kitchens one battery one battalion all of National Guard artillery and then scooped up the colonel and the command post of the 69 Regular artillery AA (antiaircraft). In galloping over a wire fence the man next me got a bad fall but I was having such a swell time I never saw the fence which was low and my horse jumped it all right. The colonel of the 6th was very mad and refused to surrender to Capt Doyle till I came up and stuck my white pistol in his face then he was very quiet especially as I paroled him as I had no men to guard prisoners.

Major General George Van Horn Mosely, the Third Army Commander, watched the maneuvers carefully and concluded that the horse cavalry had demonstrated its continued usefulness for close-in reconnaissance.

The critique came as no surprise to Patton.

He had faith in his men. After all, George S. Patton had personally trained them.

He returned to Fort Clark, and, he said, "drank beer till my teeth floated." He washed the dust and grime from his throat and silently thirsted for war. Europe was so far away. Maybe, he even hoped, hostilities would break out with Mexico along the border again. He was a soldier. And in time of peace, he sometimes thought as he walked along the banks of Las Moras Creek, he was simply wasting his time.

He sat down in his quarters at Fort Clark and wrote of his concerns to Major General Daniel Van Voorhis at Fort Knox, Kentucky:

> We had a great war in the Third Army Maneuvers and on the last day got right back of the enemy and into his gun positions. It was great sport and the funny thing was to see the utter surprise of the enemy. They had so absorbed the bull butting tactics of the World War that they forgot they had to keep their pants buttoned or else get buggared. The more people decry Cavalry, horse or mechanized, the more we will bust them up next time. The thing we want is to retain our mobility for the last ten miles.
>
> You can count on me to keep the torch burning . . .
>
> The only out about Fort Clark is the heat but I suppose it is no worse than Brownsville and you survived that. Anyhow, it is swell for the figure; mine has dwindled perceptibly . . .
>
> Some times I almost think that there will be something doing here (along the border) in a little while. Of course I have hoped so more or less ever since 1911 and it has only happened once, but now again I think there may be a flare up. If you command the army of occupation don't forget to take the Fifth Horse and loan us a few combat cars; anything you can spare.

George S. Patton longed for war.

General Jonathan Wainwright didn't. He fully believed that he was destined to spend the remainder of his military years at Fort Clark, perhaps even retire there. He had a grand swimming pool built down beside Las Moras Springs, down where the Kinney County merchant used to fill his tubs, then haul them by mule to downtown Brackettville to sell the cool, clear water for a dollar a gallon. Wainwright was content to make life easy, at least bearable, for himself and the men of Fort Clark.

Wars and rumors of war changed all of that.

General Wainwright, then Patton, were ordered to Fort Myer, Virginia. After the bombs descended in death upon Pearl Harbor, Wainwright would join General Douglas MacArthur in the South

The Coming of Age

Pacific, and George S. Patton would trade his horse for a tank and bulldoze his way into the heart of Germany.

Men were standing tall. Men were dying. It was war. It was what Patton had been waiting for.

Shortly before he left Fort Clark, Patton had been rated "Superior" by General Joyce,

> ...an outstanding leader who has great mental and physical energy. Because of his innate dash and great physical courage and endurance he is a cavalry officer from whom extraordinary feats might be expected in war. A deep military student who is intensely interested in his profession. He is thoroughly qualified for the grade of brigadier general. Of outstanding value to the service in every way.

He had earned his star.

And his eyes turned eagerly toward Europe. The battles raged on, but Patton was not a part of them. He was a maverick. His obstinate opinions — always brash, always voiced too loudly and usually in the wrong place at the wrong time — kept him in constant hot water with the military brass. When word was received that an assassination attempt had been made on Adolph Hitler, Patton was horrified.

He immediately called on General Omar Bradley and said,

> For God's sake, Brad, you've got to get me into this fight before the war is over. You know I'm in the doghouse, and I'm very apt to die there, unless I have a chance to do something spectacular.

Bradley gave him his chance.

Patton's tanks roared madly across France. He had once predicted:

> Some day I'm going to bust loose across Europe and be heading hell-bent for Berlin. Then either some coward or some dirty politician is going to get worried and order me to stop. So you know what? If I ever break loose, I mean when I do, I'm going to smash my command radio receivers or else send back no position reports and keep going. Then nobody can say I didn't follow orders.

He had, at last, been unshackled and was on his way.

Sure enough, the orders came:

"Patton, you have reached your designated objective. Stop there and await further orders."

He stomped the Normandy soil, and he fumed. And Patton snapped,

> Why hell, the only thing for an army to do when it has the enemy on the run is to keep going until it runs out of gas, and then continue on foot, to keep killing until it runs out of ammunition and then go on killing with bayonets and rifle butts.

George S. Patton, some officers swore, was intent on winning the war by himself.

Sometimes he needed help.

The rains and a gray, covert fog hung above the Ardennes Forest, forcing the allied planes from the sky as a German force hammered its way fifty-two miles into Belgium, threatening to cut the allied armies in half. Von Rundstedt must be stopped. But the damp fogs hid him, and Patton groped blindly in pursuit.

For hours he had stood staring out into the darkness that defied him. Suddenly Patton turned to a Third Army chaplain and said, "I want a prayer to stop this rain. If we get a couple of clear days, we could get in there and kill a couple of hundred thousand of these krauts."

The chaplain was dumbfounded. "It is not in the realm of theology to pray for something to help kill your fellow man," he argued.

Patton's face reddened. "What the hell are you?" he demanded to know, "A theologian or an officer of the United States Third Army. I want that prayer."

He got it. And he prayed:

> Almighty and most merciful Father, we humbly beseech Thee, of Thy great goodness, to restrain these immoderate rains with which we have had to contend. Grant us fair weather for battle. Graciously hearken to us as soldiers who call upon Thee that armed with Thy power, we may advance from victory to victory and crush the oppression and wickedness of our enemies, and establish Thy justice among men and nations. Amen.

By morning, the rains had dried from the sky.

And Patton's tanks smashed Von Rundstedt's triumphant charge through Belgium.

General Bradley later said,

> I had some genuine reservations at the thought of having George in my command. Knowing his temperament, I was afraid that he would be terribly disappointed and thus bitter and resentful. Instead, he acted like a real soldier.

And George S. Patton bullied his way into the war-torn innards of Germany itself. It was a land both disorganized and disoriented. But for the moment, the land and the glory belonged to him.

He promptly wired to the Commanding General of ComZone in Paris: "Have just pissed in Rhine, for God's sake send gasoline."

For General Jonathan Wainwright, the journey from Fort Clark was long and arduous. For him, war began and it ended in the torment, the insanity of Bataan, of Corregidor. As he later recalled,

> Jap soldiers fought with fine guns, grenades, artillery, and planes, the very best they could copy from other nations. But mentally and morally they were fighting with spears, pikes, and flaming arrows. The banzai charge, the ritualistic harakiri, the bizarre kamikaze were not the tactics of reason but the processes of people and leaders with only a remote appreciation of man's stature in the twentieth century. The Japs' horror of surrender and defeat sprang from nothing more substantial than a foggy, medieval belief that the victors would reduce them to common slavery.

Jonathan Wainwright became their slave.

He survived the death march.

He survived — period.

And when the general returned to the United States after being freed from his prisoner of war camp, he stood before 100,000 people beside the Washington Monument, and, with tears in his eyes, told them:

> We lived in a blacked-out world during our prison days. One of the least, and the greatest, of the cruelties practiced by the Japs was to keep us from frequent contact with home. We seldom knew what had happened to our loved ones.
>
> From the poverty of our existence out there, we have returned to find America strong and great. Even before we first set foot on the American continent at San Francisco last Saturday we knew how this country had rallied from our defeat at Corregidor.
>
> We saw the strong, seasoned American troops who had defeated the Japanese in campaign after campaign. We saw the wealth of air power in great planes which were hardly blueprints in the days when we anxiously scanned the skies for the relief that did not exist. We saw the mighty naval armada, risen from the grave of Pearl Harbor, stretched out across the waters of the Pacific to menace the now cringing Japs. The power of America was assembled out there, and we thanked God for it.
>
> The men who fought on Bataan and Corregidor were never beaten in spirit. Exhausted by thinning supply and the ordeal of terrific pounding by siege guns and bombers, it was useless to continue the struggle. We surrendered as honorable soldiers.

You know what happened after that. The rights and privileges which civilized nations have agreed to grant prisoners of war were denied by the Japs. Many brave and gallant soldiers died under the torment and starvation they were forced senselessly to suffer . . .

Nothing can restore the men who died for their loved ones. Yet their sacrifice, living on in the thoughts and deeds of America, can protect this nation from the lack of practical foresight which brought about those tragic events.

As I stood on the deck of the *Missouri*, at the right hand of General MacArthur, watching the signing of the surrender document, I fervently wished every American could feel the full significance of that moment. Nearly four years had elapsed since the Japs launched their attack on Pearl Harbor and on the Philippines.

That moment of surrender in Tokyo Bay had been bought with the blood of more than a million Americans who died or were wounded in the struggle. Billions of dollars and countless hours of work by Americans at home had been required to bring that party of beaten Japs to the *Missouri*'s deck . . .

It is over now, and we are at peace . . .

That afternoon in the Rose Garden of the White House, General Wainwright watched as President Harry S Truman stepped before a battery of microphones and began to read:

General Jonathan M. Wainwright, commanding United States Army forces in the Philippines from March 12 to May 7, 1942, distinguished himself by intrepid and determined leadership against greatly superior enemy forces. At the repeated risk of life above and beyond the call of duty in his position, he frequented the firing line of his troops where his presence provided the example and incentive that helped make the gallant efforts of these men possible.

The final stand on beleaguered Corregidor, for which he was in an important measure personally responsible, commanded the admiration of the nation's Allies. It reflected the high morale of American arms in the face of overwhelming odds. His courage and resolution were a vitally needed inspiration to the then sorely pressured freedom-loving peoples of the world.

And so it gives me more pleasure than almost anything I've ever done to present General Wainwright with the Congressional Medal of Honor — the highest honor this country can bestow on a man.

Those long, torturous nights — locked away behind barbed wire and facing the bayonets of cruel jailers —were behind him.

Jonathan M. Wainwright was home.

10

Epilogue

The troops at Fort Clark had prepared for a world conflict long before Hitler ever marched into Poland, long before the Japanese bombers dropped out of the rising sun above Pearl Harbor. As early as 1936, the First, Fifth, Seventh, Eighth, and Twelfth Cavalry Divisions from Forts Clark and Ringgold had gotten together for war games out in the desert lands around Marfa. As one sergeant remembered, "We would all mount up and take off enmasse down the road, and it would be so dusty that you couldn't see the man next to you."

Sometimes the men rode thirty-five miles a day. Sometimes they went further. It all depended on where the pools of water lay, and sometimes there was no water at all.

The sergeant recalled, "The longest ride I ever made in one saddle sitting was eighty miles. We could have ridden longer, but there wasn't any sense in killing good horses."

The skirmishes, the war games lasted for six months before the cavalrymen got back home to their barracks at Fort Clark again.

Three years later, as the thunder of war rolled louder above Europe, the Cavalry rode to Balmorhea on the dust storm fringe of West Texas. Fort Clark was left with only enough men to keep the post secure and feed the animals. For months, the cavalrymen lived in tents, fighting their mock war, growing thirsty and disgruntled when the big springs at Balmorhea suddenly dried up. The sergeant later said, "When our bread was too hard, we'd just soak it up with gravy."

On horseback, the troopers trained diligently for actual combat. Their tactics varied little from the days when John Bullis and his Seminole Scouts had driven the Indians off the ground that now held their tents. Time, however, had passed them all by.

When the Fifth Cavalry finally went to war, the men marched away from Fort Clark as foot soldiers.

The post would never be quite the same again. The *esprit de corps* just sort of drifted away from Las Moras Creek. For so many, an assignment to Fort Clark was like being banished to the outer reaches of Siberia. It was punishment, the last duty station for anyone who had ever done anything wrong in the Army. For some, it was purgatory, an out-of-the-way place where the Army expelled its wayward the misfits, the trouble makers. There was little authority at Fort Clark, hardly any discipline, and no respect at all.

One Colonel had a little black dog who always ran down the bridle paths in front of him. The men kept a sharp lookout for the puppy. When they saw him, they knew they had better "run like hell to get out of the way," or get ready to stand at attention and salute the brass. Saluting seemed like an awful waste of time. But sometimes, the soldiers didn't run fast or far enough.

Every time the Colonel returned a salute, he would mutter under his breath, "Same to you."

He never failed.

One afternoon, a young officer turned to him and asked, "Sir, why do you always say 'Same to you' when you salute."

The Colonel frowned and answered, "I know what they're calling me under their breath."

Fort Clark had its chaos.

The Air Corps dispatched a small unit of planes to the post, and a detail of men slowly cleared out a makeshift airfield. One of the soldiers heard that the young ladies of Brackettville were out sunbathing in the nude on top of their father's barn. That was a sight he had never seen before. That was a sight he couldn't miss. In the quiet hours of a hot afternoon, the trooper slipped out to the airfield, climbed into a general's plane, studied the control panel for a few minutes, revved up the engine, and took off — making a series of long, slow sweeps at low level above the farmer's barn.

The girls were surprised, but not nearly so much as the general.

The young soldier was almost court-martialed. But the army, in its benevolence, decided not to charge him with stealing the gener-

Epilogue

al's airplane. Instead, he was jailed for merely taking "an unauthorized flight."

And Fort Clark had its confusion.

From behind his desk, the officer of the day was just not able to see either the cannon or the flagpole, both located on the far side of the post in the middle of the parade ground. So late each afternoon, he would send a recruit out to the edge of the porch at Bullis Hall to await the officer's cue, telling him it was five o'clock and time for the Retreat ceremonies. The recruit, in turn, would wave his arm, signalling for the soldiers to lower the flag and fire the cannon. It was a well-conceived plan, but one that required concentration and split-second precision on everyone's part.

On one particular day, the trooper was standing eagerly on the porch when a huge rat suddenly bolted between his legs. Startled, the soldier leaped backward, his eyes as wide as porcelain saucers, and he raised his arms in fright.

The flag came down. The cannon blasted off its five o'clock shot. The officer of the day screamed and cursed and knew that he was in trouble before the smoke had cleared away from the parade ground.

Retreat that day came twenty-eight minutes early.

Fort Clark had its problems — at least its troopers did.

Brackettville, for the most part, took the soldier's money but didn't care that much for him. The soldier was merely a convenience, a necessary evil — especially on payday. The town's economy grew each and every time a dispirited recruit stumbled back to his barracks, hung over and broke. Even the army had little sympathy for him.

It was reminiscent of those days when Brackettville had been placed off limits. Yet a young cavalryman, his throat dry from too many hard days on a hot prairie, slipped into town anyway. Somewhere in the darkness, he was killed by Indians and his body thrown back across Las Moras Creek. The Army didn't retaliate. The Commanding Officer simply and coldly reported: "He went to town against orders. He was killed." On another occasion, a sergeant, who believed that his fortune lay behind the faces in a deck of cards, sneaked away from the fort and found a high stakes poker game in a Brackettville saloon. He died beneath the table, his cards scattered and a bullet through his head. The Commanding Officer reported: "He joined a card-playing game in Brackettville, assumed this man cheated and was killed." The Army, it seemed, had little compassion and no sentiment for those who disobeyed orders and took their lives in their own hands in the dirty streets of Brackettville.

The soldiers knew that the back streets of Brackettville were evil. The town felt the same way about the soldiers who ventured away from Fort Clark. Young ladies, on carefree evenings, could be out walking their dogs when their fathers spotted a military man coming down the sidewalk.

"Come in, Daughter," the father would yell, "and bring the dog with you."

Mistrust stalked the streets. The good, solid, God-fearing citizens of Brackettville kept their distance from the soldiers of Fort Clark. Most regarded all men in the uniform as troublemakers.

Many were.

Some weren't. Fort Clark had its heroes, too.

For so long, the Cavalry had been the heart and soul of the outpost. It could move fast, travel light, and had tamed the Texas frontier with merely a horse, a pistol, a saber, and sometimes a carbine rifle. The Cavalry, by tradition, was fearless, able to withstand any punishment during those long, forced rides into the desert.

Texas had always had its own Cavalry, known by such names as the "Prairie Rangers," the "Texas Mounted Rifles," the "Partisan Rangers," even "Bremond Light Guard," and "Whitfield's Legion." In World War II, the 112th Cavalry became the last combat trained troops of the Fifty-sixth Horse Cavalry Brigade to move into action overseas. During their days at Fort Clark, the 112th Cavalrymen joined maneuvers that led them from Fort Bliss in El Paso, all the way into Louisiana, then back to Las Moras Creek again.

One officer recalled,

> The ride had been hot and dusty and long, almost 1,500 miles. But the Cavalrymen didn't want to ride back into the fort looking worn and bedraggled. So outside of Brackettville, they stopped for a moment, dusted themselves and their saddles, then rode into Fort Clark with straight backs and their heads held high.

The 112th drove away from the prairies of Texas as an Armored Cavalry and tore across the soil of Australia, New Guinea, then New Britain in the Philippines. They didn't slow down until they fought their way into the interior of Japan. In 1943, the unit received the coveted Arrowhead on its campaign streamer for an amphibious assault landing in the Bismarck Archipelago. Fighting with the 148th Field Artillery Battalion, its aim had been to divert the Jap's attention so the First Marine Division could invade Cape Cloucester. For six months, the 112th Armored Cavalry was trapped in bitter, deadly jungle fighting, constantly punished by Japanese bombs as they struggled to hold on to, then possess, the western half of the

Epilogue

New Britain. On the Driniumor River in New Guinea, the cavalry battled the Japanese 18th Army for forty-three long, agonizing days.

In 1945, Pat Flaherty of NBC News, would broadcast the heroism of the 112th Armored Cavalry to the nation. He said:

> Thus far, the 112th Cavalry has been, more or less, an independent outfit and fought in many battles since 1942 . . . The bantam 112th Cavalry was strung out along the 5,000 yard riverfront near a village subject to heavy enemy attack. The Japs did very little fighting during the day. But night after night, they threw suicide squads whooping and yelling, 'Banzai' until the trigger-weary troopers had decimated four Japanese regiments. The battle raged on for fifty-one days of fighting. Over 2,000 Japs have been killed from a range of three yards to twenty-five meters. When the troops started staging for the invasion of the Philippines, the 112th became an active regiment of the 1st Cavalry Division. Perhaps you didn't know they were part of the Sixth Army. But the general knows it, and so do the Japs.

The men of Fort Clark had left a definite, unforgettable mark on the war in the Pacific.

At the Post in Southwest Texas, officers were in a quandary. They had long known that the military reservation at Fort Clark was much too small to hold battlefield maneuvers, yet the troops had to be trained for combat conditions. They looked out at the prairies where wars had been fought before. The land was fenced now. Cattle and horses grazed upon its short grasses, sheltered by those gnarled, crooked mesquite groves. The land was valuable all right. And the Army wanted it.

Out on his ranch, James T. Shahan knew there was a war going on, and it was an honorable war, and he stood with pride every time he saw Old Glory waving high above the rooftops of Brackettville, which was often. He disdained German beer and always felt he'd lost a friend when someone played taps beside a fresh-cut grave, even when those faceless soldiers lowered the tear-stained coffin of a stranger, and almost all of them were.

The fight hadn't been his at all.

Then suddenly it was, and Shahan was mad as hell as he tore across the countryside, slapping bullets into his rifle and swearing he would use them if he had to, and Shahan never rode that hard unless he had to.

He cursed softly to himself. War indeed was hell. He had been invaded and the time had come to take a stand or run, and James T. Shahan never ran, not even in the face of overwhelming odds, and he was definitely in the face of overwhelming odds.

He had stood alone before, and it didn't particularly frighten him. He was too angry to be scared. Besides, he had his rifle, and it was loaded with five shots, and that was usually all it took to put an end to any kind of disorder.

Shahan wondered if it would be enough to stop the whole damn U.S. Army.

Back in the middle of July, while mesquite sopped up the last moisture from those dry rolling hillsides, the government had come into the South Texas brush country and begun appropriating vast chunks of land where its fighting men could learn the deadly ways of war. The army had laid claim to 8,000 acres of Shahan's Black Angus Ranch, ordering him to move his cattle out so the military could move its cavalry in.

That was all right.

James T. Shahan didn't mind.

It was 1943, and there was a war going on, and it was an honorable war, and pouring out his German beer had been the American thing for him to do. Men were sacrificing their lives. Shahan knew he could damn well sacrifice a little land, and he was proud to give it up.

Three days earlier, Frank Clamp had come riding at daybreak, bringing a message that wiped some of the pride from his eyes.

"Them army boys are on your land," Clamp said.

"I know," Shahan had replied. "I let 'em have it."

"They been shootin' at your water tank."

Shahan frowned. Water was a precious commodity in the Texas brush country. It was life. If he lost his water, he could lose his whole struggle for survival on a parched land that had been trying to get rid of him for years anyway.

"Did they hit it?" Shahan had asked, stepping off the wooden porch of his hilltop home.

"They flat put holes in your windmill," Clamp told him. He would have spit with disgust if his mouth hadn't been so damn dry.

Shahan had nodded and gazed toward the far range where the cavalry was camped. It must have been some fool hothead, he thought, or maybe a drunk who had quenched his thirst and drowned his good sense with the same cheap downtown whiskey. Mistakes happen. Shahan allowed the army *one* — and the army had just used it up.

He waited.

He didn't wait for long.

Two nights later, Clamp returned, dust and frustration plastered across the sweat on his face. "They're back," the cowboy said.

"They're holdin' maneuvers on your pastureland and usin' your water tank for target practice."

"We lost any water?"

"Not much. But you've flat lost a few cows."

"How come?"

"Them army boys done started cuttin' down your fences."

Shahan's eyes narrowed, and his jaw tightened as though he were chomping on a bullet, and maybe he was. "Did you tell 'em to get the hell off my land?" he asked.

"Yes, sir."

"And what'd they say?"

"The lieutenant laughed at me."

Noboby crossed Shahan. He hadn't been able to tame the land around him. Nobody ever could. But he could damn sure tame the men who dared ride across it, especially if they came riding his way. No doubts at all about who was boss. The Army would be no exception.

Shahan reined his horse to a stop and checked the .30-30 saddle rifle that an old Texas Ranger had given him. He rode slowly to the big gate that separated his pastureland from the army's reservation and watched as a cloud of ragged dust rolled closer his way. He could hear the muffled thud of the horse hooves on the sunbaked plains, and he pushed his hat back with the tip of his rifle as the lieutenant galloped toward him.

"Don't open the gate," Shahan yelled. His voice hit the young officer like a pistol shot.

"What's wrong with you?" snapped the lieutenant.

"This is my land. There's no reason for you to cross it, so just turn around and head back toward the land the government took for you. The pasture still belongs to me."

The lieutenant, given courage by the company of men behind him, stood in the stirrups and boasted, "You can't stop the army."

"Maybe not."

Shahan grinned and leveled the saddle rifle just about even with the silver bars on the lieutenant's helmet. "But I can stop you," the rancher said. "Your men may mess up my pasture and waste my water, but they're gonna have to ride over your body to get there."

Shahan was still grinning when the company of U.S. soldiers rode out of sight.

Colonel Johnson at Fort Clark was appalled. He telephoned Shahan and said loudly, "I can march my men anywhere I want, anytime I want."

"If you march'em on my land," Shahan answered softly, "bring a shovel and gravedigger."

The colonel hung up and leaned back in his chair. He knew Texas cattlemen and knew what it was like to tangle with them. Maybe he would just go ahead and send his brash-faced lieutenant to Germany, he decided, and let him fight a war that he might have a chance of winning.

During the 1940s, Fort Clark was changing. The horse had become obsolete. Needed mobility was provided by motor vehicles. The horse soldiers were a lost and vanishing breed, and long, honorable military careers were dying with them.

The numbers dwindled, and men were no longer able to climb into saddles and ride, sharp as December wind across the prairies. Troopers simply mounted and hung on, and a motley band of cavalrymen charged in ragged disarray, fearful riders atop old and sweating horses. On one afternoon, near Spofford Junction, the cavalrymen had been practicing such gaits as a walk, a trot, and a gallop, when the executive officer turned them all toward home. One lieutenant recalled, "We got up to the gallop and the men lost control of their horses, and so as we finally ended up the last quarter of a mile, we looked like Gettysburg on the third day."

There was more smoke.

More dust.

The colonel didn't settle down for days.

The glory days of the cavalry were ending. Perhaps it would never shine brightly again. A cavalry division was made up of horses, artillery, and reconnaissance vehicles. There might be a pack train of mules. There might be a pack train of trucks.

So there were strong arguments to close Fort Clark. There were pleas made to keep it open.

And they were made in vain.

Fort Clark had finally run out of time.

In 1945, the Secretary of War officially declared that Fort Clark was surplus. It wasn't wanted anymore. It wasn't needed. And a year later, the troops marched away from the lonely sentinel on Las Moras Creek for the last time. The taps, on that morning, were for the fort itself, not for the men who lay at rest within the sacred ground of its graveyard. Fort Clark stood deserted, out of place and out of touch with an army that had depended on the outpost for almost a century.

Kinney County had the priority, by law, to purchase the fort, but the Commissioner's Court, by a vote of three to two, refused to exercise its right to spend so much money to maintain a fort that had protected its land for so many years. The commissioners, simply, had no use for it. It was their opinion that the residents of the county would be better off if the military reservation were used for some "worthy public purpose." Or perhaps they thought some private individual or corporation might be able to come in and resurrect Fort Clark before it wasted away in the ashes of its own neglect.

Two bills were even quickly introduced in Congress to authorize the transfer of the facilities to the Fort Clark Military Institute, a non-profit enterprise that would be led by such directors as former Vice-President John Nance Garner and Amon G. Carter, publisher of the *Fort Worth Star-Telegram*. The proposal seemed to many like a solid, even brilliant, idea. After all, so many of the nation's soldiers had been trained for the world's battlefronts at the old outpost. Now the nation's youth would be able to walk its historic grounds in pursuit of a military education, studying in buildings where Ranald Mackenzie, Philip Sheridan, John Lapham Bullis, George S. Patton, and Jonathan M. Wainwright had plotted the course of a country at war. But, alas, the transfer never came, and a dream passed away amongst the mesquite and chaparral.

Fort Clark was turned over to the War Assets Administration for disposal. Bids were submitted as the government sought to sell the post and its facilities outright, and forever wash its hands of the fort. The highest offer came from the Texas Railway Equipment Company of Houston, a bid of $411,250. Fort Clark was stripped, and the salvage sold, some have estimated, for as much as four million dollars. Since then, it has been converted into a successful retirement community, and many of the old stone building are still standing.

For ninety-four years, Fort Clark stood firm in defense of a country whose flag flew high and stately above its groves of live oaks and pecans. It was cursed. It was the answer to a pioneer's prayer. Indians taunted the lonely sentinel. Its horse soldiers hunted them and chased them and drove them from the prairies that were stained by the blood of those who had defied the warriors and paid a terrible price. A Union Army called Fort Clark home. So did the rebels of a doomed Confederacy. Seminole Negroes were paid with honor upon its parade ground. Seminole Negroes became outcasts who stole food from garbage cans when there were no more Indians for them to track across the untrackable deserts of Texas.

From Fort Clark, men followed Teddy Roosevelt to battle the Spanish upon Cuban soil. They hunted Pancho Villa and his bandits as they stalked them through the thirsty sands, beyond the adobe walls of Mexico. They sailed away to end the war that would end all wars. They crossed the Rhine in Patton's tanks. And they found the fringes of hell in Bataan and Corregidor.

Men fought each other at Fort Clark, in mock wars and in backstreet brawls. They found love for two dollars and whiskey as cheap as love. They sometimes thought that *they* were being punished behind the barbed wire fences of the fort, not the German prisoners of war who seemed content just to be so far removed from the bullets that had caused their surrender. None of the prisoners ever escaped. None ever wanted to. They looked out across a vast, parched prairie and knew there was no place to go.

Fort Clark helped pave the way, as America sometimes marched and sometimes stumbled toward military greatness. Fort Clark became a wizened, grizzled veteran with many battle scars and many campaign ribbons. Medals of Honor hung from its neck. And pride always outshone the bitterness in its eyes.

It was a good soldier.

It fought a good fight.

Fort Clark was never given an order that it didn't carry out. And Fort Clark was destined to remain as a lone sentinel, an old soldier that had died but would never fade quietly away.

Kitchen and Mess Hall — 1870s
Courtesy National Archives

Hospital — 1870s
Courtesy National Library of Medicine
Bethesda, Maryland

Adobe Blacksmith Shop Built in 1855—1880s
Courtesy National Library of Medicine
Bethesda, Maryland

Seminole Indian Hut or Mexican *Jacales*— 1890s

Officers pose for photographer — 1880s

Interior – Barracks — 1880s

Rear View–Field and Staff Quarters — 1880s
Courtesy National Archives

Post Headquarters — 1880s
Courtesy National Archives

Post Guardhouse — 1880s
Courtesy National Archives

Post Bakery — 1880s
Courtesy National Archives

Post Hospital — 1880s
Courtesy National Archives

Cavalry Stables — 1880s
Courtesy National Archives

Las Moras Spring and Pumphouse — 1880s

Las Moras Creek — 1880s
Courtesy Library of Congress

Laundress House and Cook shack — 1870s
Courtesy National Archives

Bandstand — 1880s
Courtesy Library of Congress

Barracks—Enlisted Men — 1880s
Courtesy Library of Congress

Officers Row — 1890s
Courtesy Fort McIntosh Armory – Military
Antiques — McIntosh, Florida

On Parade — 1890s
Courtesy Fort McIntosh Armory – Military
Antiques — McIntosh, Florida

Gathering Water – Las Moras Creek — 1890s
Courtesy Fort McIntosh Armory – Military
Antiques — McIntosh, Florida

Seminole – Mexican Family — 1890s

On Parade — 1890s
Courtesy Library of Congress

On Maneuvers — 1890s
Courtesy Library of Congress

Parade Ground — 1890s
Courtesy Library of Congress

War Games — 1890s
Courtesy Library of Congress

Full Parade — 1890s
Courtesy Library of Congress

Brackettville in background — 1892
Courtesy National Archives

In need of Repair — 1900
Courtesy National Archives

Enlisted Men's barracks — 1900s
Courtesy National Archives

Officers sit for formal picture — 1902
Courtesy Pentagon

Tailor Shop — 1900s
Courtesy National Archives

Seminole Indian Scouts — Unknown, Billy July, Ben July, Unknown, Ben Wilson, John July, William Shield, 1900s
Courtesy Institute of Texan Cultures
San Antonio, Texas

Seminole Indian Scouts — 1900s

George S. Patton — 1938
Courtesy Fort Hood Museum

Seminole Scout John Jefferson — 1890s
Courtesy Institute of Texan Cultures

Colonel Ranald Mackenzie (in later life) —1880s
Courtesy National Archives

John Lapham Bullis — 1880s
Courtesy Institute of Texan Cultures

Teamsters camped at Las Moras Springs — 1880s
Courtesy Institute of Texan Cultures

On Review — 1920s
Courtesy National Archives

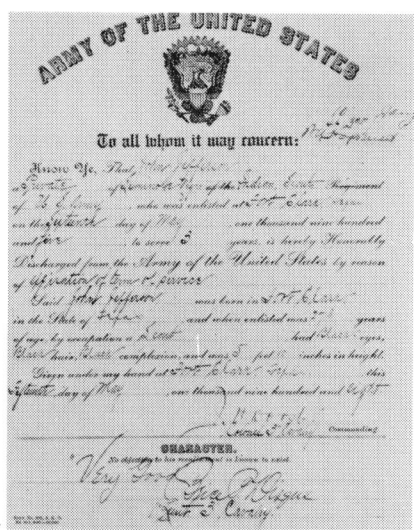

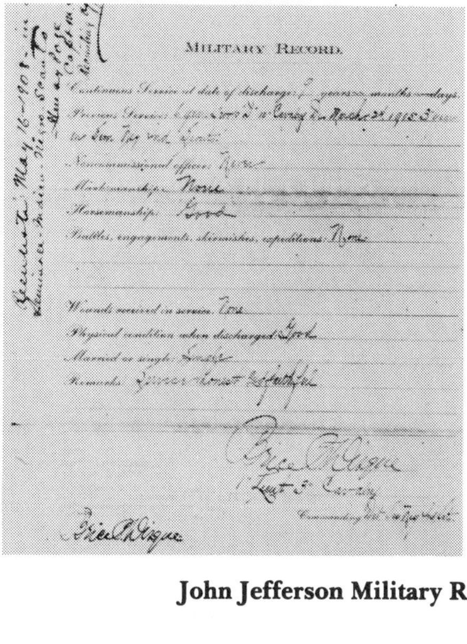

John Jefferson Military Record Report — 1908
Courtesy Institute of Texan Cultures

General Jonathan M. Wainwright and wife before leaving Fort Clark — 1930s
Courtesy Daughters of the Republic of Texas Library

Fort Clark from the air — 1929
Courtesy National Archives

Post Movie Theater — 1930s
Courtesy National Archives

Brackettville, Texas — 1930s
Courtesy Fort Clark Historical Society

Brackettville, Texas — 1920s
Courtesy The University of Texas
Humanities Research Center

Swimming Pool — 1930s
Courtesy The University of Texas
Humanities Research Center

HQ Fifth Cavalry — 1938
Courtesy Daughters of the Republic
of Texas Library

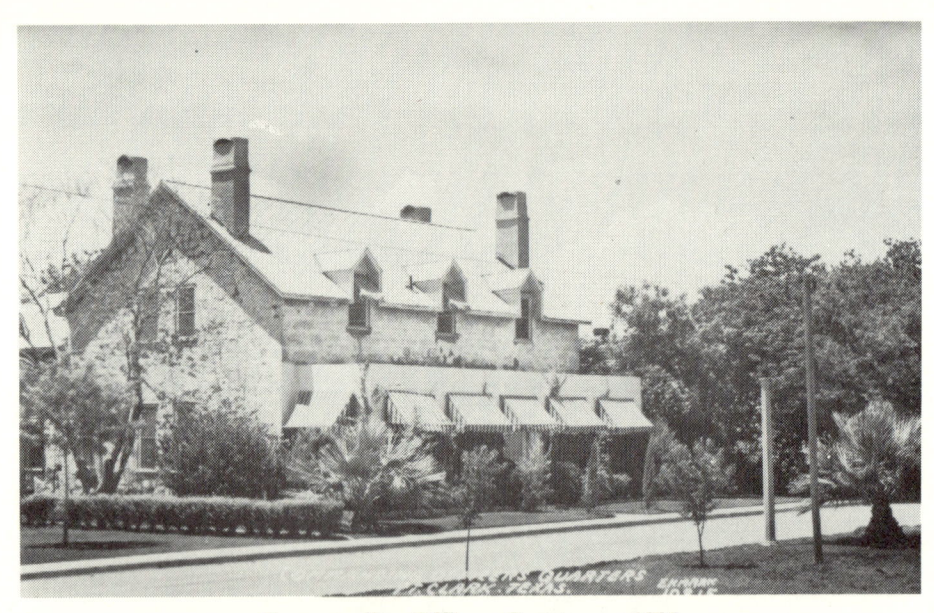

Commanding Officers Quarters — 1938
Courtesy Daughters of the Republic
of Texas Library

Fifth Cavalry Marching — 1938

Brigadier General R. C. Richardson
leaving Fort Clark — 1938
Courtesy Daughters of the Republic
of Texas Library

Officers Wives waving *Good-Bye*
to Brigadier General R. C. Richardson — 1938
Courtesy Daughters of the Republic
of Texas Library

"At Ease" — 1938
Courtesy Daughters of the Republic
of Texas Library

Cavalry Maneuvers at Tayanvale, Texas — October 1939.
Generals Wainwright, Brees, and Joyce — 1939
Courtesy Daughters of the Republic
of Texas Library

Field maneuvers — Generals Steans, Wainwright, and Robenson — 1937

Courtesy Daughter of the Republic
of Texas Library

Horse Picket Line on maneuvers — 1938

Courtesy Daughters of the Republic
of Texas Library

Blanket Toss-Up — 1938
Courtesy Daughters of the Republic
of Texas Library

On the Trail — Troop E Twelfth Cavalry —1938
Courtesy Daughters of the Republic
of Texas Library

Rain in the Desert? — 1938
Courtesy Daughters of the Republic
of Texas Library

Slip and Slide — 1938
Courtesy Daughters of the Republic
of Texas Library

Fort Clark Horse Show—Sergeant Corey, Fifth Cavalry — 1939
Courtesy Daughters of the Republic
of Texas Library

Mass Maneuvers, Fifth Cavalry —1939
Courtesy Daughters of the Republic
of Texas Library

Quartermasters Storehouse and Office — 1930s
Courtesy Daughters of the Republic
of Texas Library

Fifth Cavalry, Firing Range — 1930s
Courtesy Fort Clark Historical Society

On Parade — 1940
Courtesy Daughters of the Republic
of Texas Library

**Brigadier General Wainwright and Staff,
Balmorhea Maneuvers — 1939**
Courtesy Daughters of the Republic
of Texas Library

Wooden Enlisted Men's Barracks — 1941
Courtesy National Archives

Post Chapel — 1944
Courtesy Brown and Root

Officer's Club — 1940
Courtesy National Archives

Major General Kenyon A. Joyce — 1939
Courtesy National Archives

General Jonathan M. Wainwright — 1930s
Courtesy Pentagon

Post Headquarters — 1942
Courtesy Fort Clark Historical Society

Post Entrance — 1943
Courtesy Fort Clark Historical Society

Troop Barracks — 1942
Courtesy Fort Clark Historical Society

Officers Wives and Children on lawn — 1942
Courtesy Fort Clark Historical Society

Laundry Building — 1944
Courtesy Fort Clark Historical Society

Tar Paper Building — 1944
Courtesy Fort Clark Historical Society

Sergeant L. B. Walker instructing troops —
Courtesy Pentagon

Troops unloading at Spofford, Texas — 1943
Courtesy Pentagon

Troops unloading at Spofford, Texas — 1943

Courtesy Pentagon

Taking a Coke Break — 1943

Courtesy Pentagon

Bibliography

Books

Anderson, John Q. *Tales of Frontier Texas*. Dallas, Texas: SMU Press, 1966.

Ayer, Fred, Jr. *Before The Colors Fade*. Dunwoody, Georgia; Norman S. Berg, Publisher, 1971.

Bluemenson, Martin. *The Patton Papers, 1884–1940*.

Boyd, Mrs. Orsemus. *Cavalry Life in Tent and Field*. New York, 1894.

Carter, Captain R. G. *On The Border with Mackenzie*. Washington, D.C.: Eynon Printing Company, Inc., 1935.

Crane, Charles Judson. *The Experience of A Colonel of Infantry*. New York, 1923.

DeWees, William B. *Letters From An Early Settler in Texas*. Waco, Texas: Texian Press, 1968.

Foreman, Grant. *Indians and Pioneers*. Norman, Oklahoma: University of Oklahoma Press, 1936.

———. *Indian Removal*. Norman, Oklahoma: University of Oklahoma Press, 1953.

Foster, Laurence, *Negro-Indian Relationships in the Southeast*, Philadelphia, 1935.

Giddings, Joshua. *The Exiles of Florida*. Columbus, Ohio: Follett Foster and Company, 1859.

Grant, Ulysses, S. *Personal Memoirs of U.S. Grant*, two volumes. New York, 1886.

Hill, Luther B. *A History of The State of Oklahoma*, Vol. 1. Chicago and New York: The Lewis Publishing Co., 1909.

Holt, Roy D. *Heap Many Texas Chiefs*. San Antonio, Texas: The Naylor Company, 1966.

Horgan, Paul, *The Great River: The Rio Grande in North American History*, New York, Rhinehart & Company, 1954.

Jahoda, Gloria. *The Trail of Tears*. New York: Holt, Rinehart & Winston, 1975.

James, Vinton Lee. *Frontier and Pioneer Recollections of Early Days in San Antonio and West Texas*. San Antonio, Texas: Artes Graficas, 1938.

Kimball, Maria Brace. *A Soldier–Doctor of Our Army, James P. Kimball*. Boston: Houghton–Mifflin, 1917.

Kinney County Historical Commission, Kinney County, *125 Years of Growth*. Brackettville, Texas, 1977.

Lane, Lydia Spencer. *I Married A Soldier*. Philadelphia: J. P. Lippincott Co., 1893.

Mahon, John K. *Reminiscences of The Second Seminole War by John Bemrose*. Gainesville, Florida: University of Florida Press, 1966.

Madison, Virginia, and Stillwell, Hallie. *How Come It's Called That.* New York, New York: October House, Inc., 1968.

Mellor, William Bancroft. *Patton: Fighting Man.* New York: G. P. Putnam's Sons, 1946.

McConnell, H. H. *Five Years a Cavalryman.*

Patton, George S., Jr. *War as I Knew It.* Boston: Houghton–Mifflin Co., 1947.

Raht, Carlyle Graham. *The Romance of The Davis Mountains and Big Bend Country,* El Paso, Texas, 1919.

Rippy, J. Fred. *The United States and Mexico.* New York, 1926.

Rodenbough & Haskins, *The Army of the U. S.,* 1798–1896.

Rister, Carl Coke. *The Southwestern Frontier, 1865–1881.* Cleveland, Ohio, 1928.

Shipman, Mrs. O. L. *Taming The Big Bend, A History of The Extreme Western Portion of Texas from Fort Clark to El Paso.*

Strong, Henry W. *My Frontier Days and Indian Fights on The Plains of Texas.*

Sullivan, Capt. C. J. *Army Posts and Towns — The Baedeker of The Army, 1926.*

Tyler, Ronnie C. *The Big Bend.* Washington, D.C.: Office of Publications, National Park Service, U.S. Department of the Interior, 1975.

Vanderwerth, W. C., Compiler. *Indian Oratory.* Norman, Oklahoma. University of Oklahoma Press, 1971.

Wainwright, General Jonathan M. *General Wainwright's Story,* edited by Robert Considine. Garden City, New York: Doubleday & Co., 1946.

Wallace, Ernest (editor). *Ranald S. Mackenzie's Official Correspondence Relating to Texas, 1873–1879,* two volumes. Lubbock, Texas: West Texas Museum Association, 1968.

Walton, W. M. *Life & Adventures of Ben Thompson.* Austin, Texas: Edwards & Church, publishers.

Webb, Walter Prescott, *The Great Plains,* Boston, Ginn & Company, 1931.

Winfrey, Dorman H. *Frontier Forts of Texas.* Waco, Texas: Texian Press, 1966.

Wilbarger, J. W. *Indian Depredations in Texas.* Austin, Texas: The Steck Co., 1935.

Periodicals

Albee, Capt. George E. "The Battle of Palo Duro Canyon," *The New York Herald,* October 16, 1874.

Bateman, Chaplain Cephas C. "Old Fort Clark, A Frontier Post," *Hunter's Frontier Times,* April, 1925.

Bonnett, W. A. "King Fisher," *Frontier Times,* Vol. III.

Butler, Grace Lowe. "General Bullis: Friend of The Frontier," *Frontier Times,* Vol. XII.

Harper, Walter, "Twenty-One Days Between Rivers," *Frontier Times,* October–November, 1974.

Jones, H. Conger. "Seminole Scouts Still Thrive on Border," Val Verde Historical Commission.

Kemp, Benjamin. "Feather Fort," *Frontier Times*, April, 1979.
McCright, Grady E. "John Bullis," *True West*, October, 1981.
Myer, Albert J., "Letters from Texas, 1854–1856," *Southwestern Historical Quarterly*, Vol. LXXXII, July, 1978.
Oates, Stephen B. "John 'RIP' Ford, Prudent Cavalryman, C.S.A.," *The Southwestern Historical Quarterly*, January, 1961.
Porter, Kenneth Wiggins. "The Seminole in Mexico," *The Hispanic American Historical Review*, February, 1951.
———. "The Seminole Negro–Indian Scouts, 1870–1881," The Southwestern Historical Quarterly.
Salter, Bill, "Negro Seminole Scouts Buried in Cemetery near Brackettville," *San Antonio Express News*, June 8, 1969.
Scheips, Paul J. "Albert James Myer, An Army Doctor in Texas," *The Southwestern Historical Quarterly*, July, 1978.
Schneider, George A. (editor). "A Border Incident of 1878 from The Journal of Captain John S. McNaught," *The Southwestern Historical Quarterly*, October, 1966.
Wallace, Edward S. "General Ranald Slidell Mackenzie, Indian Fighting Cavalryman," *The Southwestern Historical Quarterly*.
———. "Border Warrior," *American Heritage*, June, 1958.
Woodhull, Frost. "The Seminole Scouts on The Border," *Frontier Times*, December, 1937.
"Texas Border Troubles," *House Miscellaneous Documents*, 45th Congress, 2d Session, VI, No. 64, No. 13, No. 275.
"Statement Concerning Fort Clark, Texas," *The Texas Collection*, Baylor University, March 26, 1940.
Resolution of Texas Legislature Commending Lt. J. L. Bullis, April 7, 1882.

* * *

Complete information on McConnell, H. H., "Five Years A Cavalryman," Jacksboro, J. N. Rogers & Company, Printers, 1889.

Military Reports

Report of Commissioner of Indian Affairs by Thomas L. Judge, August 26, 1844.
Report on Topography & Disease of Fort McIntosh by Assistant Surgeon Glover Perin, 1852.
Report on Topography & Disease of Camp J. E. Johnston by Assistant Surgeon Ebenezer Swift, 1852.
Report on Topography & Disease of Fort Brown by Surgeon S. P. Moore, 1853.
W. G. Freeman's Report on The Eighth Military Department, 1853.
Annual Report of the War Department, 1853–1940.

Report on Topography & Disease of Ringgold Barracks by Assistant Surgeon Israel Moses, 1854.
Sanitary Report of Fort Clark by Assistant Surgeon Basil Norris, September, 1856.
Inspection Report of Fort Clark by Colonel J. K. F. Mansfield, June 3, 1856.
Special Order 12, Department of Texas, January, 1856.
Special Order 39, Department of Texas, April 30, 1858.
Circular No. 4, Surgeon General's Office, 1870.
Annual Report of the Commissioner of Indian Affairs to the Secretary of the Interior, Washington, D.C., 1870.
Report from Edward W. Hinks, Lieutenant Colonel, 25th Infantry, to Lieutenant C. N. Gray, Post Adjutant at Fort Clark, August 26, 1870.
Special Order 123, Department of Texas, June 16, 1871.
Report of D. E. Barr to Alexander Bishop, Episcopal Bishop of Texas, May 10, 1871.
Documents 4998, 1872; 3827, 1873; 725, 1871; 3415, 1872; 1205, 1874; 2433, 1876 in Old Files, Adjutant General's Office.
Report of Inspector General's Office, 1872.
General Order No. 17, Department of Texas, November 8, 1871.
General Order No. 10, Department of Texas, May 12, 1875.
Description of Military Posts, Reservations, and National Cemeteries, Bureau of Army Records, Washington, D. C.
Circular No. 8, Surgeon General's Office, 1872.
Petition of Elijah Daniel, Fort Clark, June 28, 1873.
Colonel Edmund Schriver's Inspector General's Report on Military Posts in Texas, November 1872–January, 1873.
Report by H. M. Atkinson, Special Indian Commissioner at Fort Duncan to Edward P. Smith, Commissioner of Indian Affairs, Washington, D.C., November 16, 1874.
Circular No. 8, Surgeon General's Office, 1875. Description of Military Posts, Fort Clark, by Acting Assistant Surgeon Donald Jackson and Assistant Surgeon P. Middleton.
Circular No. 8, Surgeon General's Office, 1875. Description of Military Posts, Fort Clark, by Acting Assistant Surgeon Donald Jackson and Assistant Surgeon P. Middleton.
Outline Description of U.S. Military Posts and Stations, War Department, Quartermaster General's Office, Washington, D.C., 1875.
Report from John L. Bullis, First Lieutenant, 24th Infantry, Commanding Scout, to Lieutenant G. W. Smith, 9th Cavalry, Post Adjutant at Fort Clark.
Report to Headquarter's Department of Texas from Lieutenant General Phil Sheridan, 1878.
Report to Adjutant General of U.S. Army in Washington, D.C., by J. M. Schofield, Major General.

Bibliography

Resolution of Texas Legislature commending Lieutenant John L. Bullis, April 7, 1882.
Inspector's Report of Fort Clark, Texas, September, 1883.
Inspector's Report of Fort Clark by Lieutenant General Phil Sheridan, April 7, 1883.
Statement of the Military Service of Ranald Mackenzie, compiled by the records of the Adjutant General's office, Headquarters of the Army, March 3, 1884.
Report of Annual Inspection of Public Buildings at Fort Clark, Texas, 1884.
Military Reservation Books, Judge Advocate General's Office, editions, 1898, 1904, 1910, 1916.
Outline Description of U.S. Posts, etc., Quartermaster General's Office, 1904.
General Orders 19 and 76, Adjutant General's Office, 1904.
General Order 8, Adjutant General's Office, 1909.
Report of Operations of Punitive Expedition, to Commanding General, Southern Department, Fort Sam Houston, by Colonel F. W. Sibley, 14th Cavalry, June 13, 1916.
Report to Commanding General, Eighth Corps Area, Fort Sam Houston, Texas, from H. H. Smith, Major Medical Corps, January 24, 1927.
Survey of War Department Activities & Installations, to Commanding General, Eighth Corps Area, Fort Sam Houston, Texas, by Colonel George L. Hicks, A.G.D., January 14, 1928.
Survey of Fort Clark by Lieutenant Colonel G. Kent, I.G.D., report made to Commanding General, Eighth Corps Area, Fort Sam Houston, Texas, 1930.
Report on History of Fort Clark, made by Bureau of Army Records, Washington, D.C. to Private Hugh Hunter, Feature Editor of "Centaur," 112th Cavalry at Fort Clark.

Letters

From M. L. Courtney, First Lieutenant, 25th Infantry to First Lieutenant C. N. Gray, Post Adjutant, December 8, 1840.
From J. Milton Thompson, First Lieutenant, 24th Infantry, to First Lieutenant C. N. Gray, Post Adjutant, December 9, 1840.
From Colonel Albert Sidney Johnston, 2nd Cavalry, 1856.
From General D. C. Buell to Fort Chadbourne, 1856.
From Captain J. H. Ralston, A.G.M., to Colonel G. Croghan, March 6, 1847.
From Captain James Oakes, 2nd Cavalry, to Commander of Fort Clark, 1857.
From D. A. Ward, Captain, 25th Infantry, to Lieutenant C. N. Gray, Post Adjutant, September 8, 1870.

From Michael Cooney, Captain, 9th Cavalry, to Post Adjutant at Fort Clark, November 28, 1872.
Letters on The Mackenzie Raid into Mexico, Eugene C. Barker Texas History Center, May 27, 1873 to February 13, 1874.
From H. W. Lawton, First Lieutenant, 4th Cavalry, to General S. B. Holabird, 1878.
From Lieutenant John L. Bullis to Post Adjutant, 4th Cavalry, November 20, 1878.
From Mary Maverick to General Thomas M. Vincent, Assistant A.G., Department of Texas, 1878.
From citizens of Kinney County to S. B. Maxey, U.S. Senate, 1882.
From Ranald Mackenzie to Adjutant General of U.S. Army, 1883.
From Lieutenant General Phil Sheridan to Brigadier General R. C. Drum, Adjutant General, General of the Army, Washington, D.C., April 7, 1883.
From Brigadier General Ranald Mackenzie to Lieutenant General Phil Sheridan, November 17, 1883.
From Jasper Ewing Brady to *New York Herald Tribune*, 1889.
From William Zimmerman, Jr., Asst. Commissioner, Office of Indian Affairs, to Charles L. South, House of Representatives, June 15, 1938.

Additional Information

Personal Interview with George Wyrick, October 26, 1980.
Speech by Dr. Robert Drummond, entitled, "6 Months with The 2nd Cavalry at Fort Clark — 1943" to Fort Clark Historical Society.
Speech by Jo Kent to Fort Clark Historical Society.
Speech by Lieutenant General Samuel Myers to Fort Clark Historical Society.
Speech by Martha Blackwell to Fort Clark Historical Society.
Speech by Miss Elsie Sauer, Mrs. Adele Augur, and Mrs. J. Lee Ballantyne to Fort Clark Historical Society.
NBC News Broadcast made by correspondent Pat Flaherty, 1945.

INDEX

A

Adkisson, William, 8
Adobe Walls, 101
Alamogordo, New Mexico, 77
Alexander, James, (and son), 5
Alpine, 133
Apaches, 68, 111, 114, 116
Appomattox, 56, 77
Aransas Bay, 25
Arapahoe Indians, 106
Ardennes Forest, 160
Atkinson, Henry M., 92, 93,
Augur, General C. C., 75, 78, 93, 94, 98, 99, 100

B

"Badhand," (Mackenzie), 77
Balmorhea, 163, 164, photo: maneuvers–204
Barr, Chaplain, 59, 60
Bataan, 161, 172
Battle of Nueces, 50
Baylor, Lieutenant Colonel John R., 48, 51
Beale, Dolores, 25
  Dr. John Charles, 25, 26, 27
Beatty [a surveyor], 8
Beaumont, Colonel Eugene, 81, 86, 104
Beaver Lake, 65, 111
Belknap, General, 79
Bell, William, 5
Big Bend, 132, 133, 135, 136
Big Tree [Kiowa Chieftain], 100
Biles, 44
Birch, Jems, 44,
Black Hawk War, 44
Bliss, Zenas R., 17, 71, 72
Bloody Bend, [See Big Bend]
Blue Goose, The, (Saloon), 128
Boehm, Captain, 103, 104
Bone Springs, 135

Boole, Mrs., 128
Boquillas, 133, 134, 135, 136, 137, 138
Bowlegs, Sergeant David, 117
Boyd, Captain Orsemus Bronson, 63
  [Mrs.], 63
Brackett, Oscar B., 20
Brackettville, 43, 44, 55, 58, 59, 78, 81, 88, 89, 98, 108, 109, 123, 125, 126, 127, 128, 129, 140, 144, 145, 149, 164, 165, 166, 167, photo–185
Bradley, General Omar, 159, 160
Brady, Jasper Ewing, 127, 128
Brazos Island, 56
Brazos River, 5, 48
Brees, General, photo–198
Bremond Light Guard, 166
Brickhouse, Mary Douglas (Read), iv
Brown, Private George, 95
  Major J. M., 63
  Private Michael, 130
Brownsville, 108, 133, 158
Bruner, Jim, 97
Buell, D. C., 37
Buffalo Springs, 57
Bullis, Lieutenant John Lapham, 75, 76, 81, 87, 97, 98, 109, 110, 111, 112, 113, 114, 115, 116, 117, 118, 119, 120, 127, 164, 171, photos–190, 193
Bullis's Crossing (ford), 113
Burleson, Captain Ed, 11

C

Caddo Indians, 5
California Springs, 74
Camp Bullis, 152

Camp Eagle Pass, 142
Camp Stockton, 49
Camp Wood, 49
Canadian River, 102
Cano, Chico, 133
Cardenas, 155
Carmen Mountains, 116
Carrannista, 137
Carter, Amon G., 171
    Captain R. G., 78, 79, 81, 82, 83, 84, 85, 86, 101, 105, 106
Cerro Blanca, 138
Cherokee Indians, 5, 70
Cheyenne Indians, 100, 101, 106
Chihuahuan Desert, 113, 132
Chisos Mountains, 132
Choate, Acting Surgeon Rufus, 105
Clamp, Frank, 168
Clark, Major John B., 14, 16
Cline, 148
Clinch, General, 69
Coacoochee (The Wildcat), 69, 70
Coahuila, Mexico, 90, 114
Coffee's Trading House, 4
Coffin Stage Line, 128
Coleman, Captain, 5
Collins, Lieutenant James L., 133
Colorado River,
Colquitt, Governor Oscar B., 133
Columbus, 8
Comanche Indians, 1, 2, 3, 5, 7, 10, 11, 12, 13, 18, 20, 27, 31, 32, 33, 35, 36, 39, 42, 45, 55, 58, 66, 67, 68, 71, 73, 77, 78, 90, 94, 95, 96, 99, 100, 101, 102, 103, 104, 105, 106, 111, 112, 121, 127, 131
Comancheros, 27, 117
Compton, C. G., 134
    son, 134
    Tommy, 134
ComZone, Commanding General, 161
Congressional Medals of Honor, 106, 108, 112, 129, 162, 172
Congressional Medals rescinded, 129

Cooney, Captain Michael, 73
Cope Ranch, 73
Corbin, Brevet Lieutenant Colonel H. C., 60
Corey, Sergeant, photo–202
Corregidor, 161, 162, 172
Corrigan, Private Peter, 88
Corro Bianca, 137
Cortina, Juan Nepomuceno, 49
Costilietos, (Chief of Lipans), 84
Cox, Ike, 68, 80, 84
Cramer, Lieutenant, 137, 138
Creek Indians, 70, 72
Croghan, G., 10
Cusack, First Lieutenant Patrick, 73, 74

D

Daniel, Elijah (Chief), 72, 97
    Private, 130
Davidson, Lieutenant, 68
Davis, Lamar, 133
Davis Mountains, 113
Deemer, Jesse, 134, 135, 136, 137, 138
Degener, Edward, 50
Del Rio, 88, 89, 140, 143,
Department of Texas (Army), 113
de Rabago, Captain Philipe,
Devil's River, 40, 65, 67, 73, 74, 116, 119
Devil's River Campaign of 1856, 39
DeWees, William B., 2, 8
Dhanis, 45
Dine, Private George, 130
Dolan Falls,
Dolores Settlement, [See Villa de Dolores], 15, 27
Donovan, Bill, 90
Doyle, Captain, 157
Duff, James M., 50
Durlon, Charles, 94

E

Eagle Pass, 18, 33, 70, 88, 107, 114, 127, 140, 142, 144
Eagle's Nest Crossing, 112
Earp, Wyatt, 126
Eddy, Lieutenant, 40

Index

Edwards Plateau, 109
Eighth Cavalry, 116, 135, 137, 138, 163
Eighth Military Department, 20
Ellis, W. K., 134
El Paso, 14, 35, 44, 74, 133
El Pino, 136, 137, 138
Elting, Commanding General Leroy, 141, 144
Everglades, [Florida], 69

F

Factor, Private Pompey, 111, 112, 113
Fifth Cavalry, 22, 142, 147, 152, 153, 157, 163, 164, photos– 196, 202, 203
Fifth Horse, 158
Fifty-sixth Horse Cavalry, 166
Fighting Third, 132
First Cavalry, 154, 163
First Cavalry Division, 167
First Infantry, 14, 16, 19, 20, 28
Fisher, King, 107, 108, 126
Flaherty, Pat, 167
Flores, Captain, 13
Florida, 68, 69, 70, 72, 97
Florida Seminole War
Ford, John Salmon (Rip), 48, 49, 51, 56
Fort Bliss, 49, 166
Fort Bragg, 148
Fort Brown, 28, 47
Fort Clark [entire book]
Fort Concho, 77
Fort Davis, 35, 49, 73
Fort Duncan, 18, 41, 47, 71, 72, 81, 99, 103, 108, 123
Fort Gibson, 92
Fort Griffin, 101
Fort Inge, 23, 47, 49
Fort Knox, Kentucky, 158
Fort Lancaster, 47, 49
Fort Martin Scott, 36
Fort Mason, 47
Fort McIntosh, 28, 49
Fort Myer, Virginia, 158
Fort Richardson, 103
Fort Riley

Fort Ringgold, 28, 29, 30, 47, 158, 163
Fort Sam Houston, 115, 144
Fort San Antonio, 10, 37, 41, 44, 51, 124
Fort Sill, 101, 106
Fort Stanton Reservation, 118
Fort Stockton, 49
Fort Sumter, 48
Fort Terrett, 58
Forty-first Infantry, 58, 76
Fourteenth Cavalry, 134, 136
Fourth Cavalry, 57, 63, 77, 78, 80, 83, 85, 87, 90, 98, 99, 101, 102, 103, 105, 106, 135
Fredericksburg, 50
Freeman, W. G., 20
Fries, John P., 109
Frio River, 118
Fritter's, 129
Funston, General, 133, 135

G

Galveston,
Galveston, Harrisburg and San Antonio Railroad, 126
Garcia, General Guadalupe, 49
Garner, John Nance, 145, 171
Garrison, Lindley M., 133
Gaskill, Commander, 64
Gerard, Private Pierre, 130
Gettysburg, 76, 170
Gilder, John, 129
Glenn Springs, 134, 135, 137, 138, 139
Goacher [family], 3
Goethe, Captain, 137
Gonzales, Anastacio, 73
Grant, General U. S., 77, 80, 82
Gray, Major, 152
Grayson, Renty, 84, 96, 97
Gray Mule, The, [Saloon], 128
Guadalupe River, 8
Gunther, Captain, 105

H

Hagarty, Laurence, 109
Hagler, Lieutenant, 52
Hamner, Captain, 53
Hard, Henry E., 104

Harper's Ferry, 76
Harper, Walter, 127
Harris, Mrs., 27
Harvey, [family], 8
Hatch, Colonel Edward, 98
Hicks, Lieutenant Colonel George L., 141
Hill, Robert T., 132
  Major, 40
Hilton House, 129
Hitler, Adolph, 159, 163
Hoffman, Alex, 45
Horn, Mrs., 27
  Sons, 27
Hornsby, Daniel, 8
  Ruben, 8
Horse, Chief John, 71, 72, 107, 108, 113
Horse Show, photo–202
Houston, 57, 128, 145
Houston County, 7
Houston, Sam, 48, 49
Howard, Tom, 12
Howard's Creek, 111
Howard's Well, 73
Hudson, Lieutenant Charles L., 30, 87, 94, 95, 100
Hurricane Bill, [See Thompson], 103

I

International and Great Northern Railroad, 126
Ishatai [Comanche Medicine Man], 101

J

Jackson, Assistant Surgeon Donald, 60, 63
James, Vinton Lee, 109
Japanese, Eighteenth Army, 167
Jarvis, Surgeon N. S., 28
Jefferson, John (photo Seminole Negro Scout), 190, 191
Jesup, General Thomas Sidney, 69
John, (Seminole Indian), 67
Johnson's Run, 111
Johnston, Colonel Albert Sidney, 37, 40, 169
Jones, Private, 130

Joyce, General; [Kenyon A.], 159, photos–198, 206
July, Ben (Photo Seminole Negro Scouts), 189
  Billy (Photo Seminole Negro Scouts), 189
  John (Photo Seminole Negro Scouts), 189

K

Kansas plains, 126
Katemcy (Comanche Chief), 2
Kemppenberger, Private Leonard, 88
Kent, Lieutenant Colonel G., 146
Kibbitts, Sergeant John, 72, 98
Kickapoo Indians, 13, 18, 20, 58, 63, 67, 68, 78, 83, 84, 85, 87, 90, 92, 93, 94, 99, 101, 111, 121
Kickapoo Springs, 73, 94, 97, 98
Kimball, Dr., 131
  Colonel James P., 130
  Mrs., 130
King, J. H., 16
Kinney County, 108, 109, 113, 123, 124, 145, 148, 158, 170
Kinney, Henry Lawrence, 44
Kiowa Indians, 4, 94, 98, 100, 102
Kurg, Private Adolph, 130

L

La Aguaita, 136
Lafani, Private Joseph, 130
Laguna band, 72
Lamotte, Joseph Hatch, 15
Lane, Lydia Spencer, 23, 42, 44
Langford, J. O., 133
Langhorne, Major George T., 135, 136, 137, 138
Las Moras Creek [entire book]
Las Moras Mountains, 20
Lawton, Lieutenant H. W., 122
Lee, Lieutenant Colonel (later General) Robert E., 47, 48, 56, 77
Lepan, [See Lipan]
Lincoln, Abraham, 44
Linnville, 11
Lipan Indians, 12, 13, 18, 20, 21,

Index

33, 39, 40, 55, 56, 58, 67, 68, 71, 73, 78, 85, 87, 93, 95, 96, 99, 109, 114, 117, 118, 119, 121
Llano Estacado, 102, 106
Locke, William,
Lockhart, Matilda, 11
Lone Wolf [Kiowa Chieftain], 95, 100, 102
Longoria, Julian, 97
Loomis, Private, 130
Los Alimos, 137
Luck, Lieutenant, 50
Luckett, Nolan, [family], 5
    Colonel Philip N., 49

M

MacArthur, General Douglas, 158, 162
Mackenzie, Ranald [Slidell], 77, 78, 79, 80, 81, 82, 83, 84, 85, 86, 87, 89, 90, 91, 92, 93, 94, 98, 99, 100, 101, 102, 104, 105, 106, 110, 116, 117, 124, 125, 171, photo–190
Magruder, Colonel, 40
Maloney, Tom, 109
Mansfield, Colonel J. K. F., 40, 41
Marathon, 135
Marfa, 133, 163
Martin, Mr., 148
Matamoros, 72
Mauck, Major Clarence, 80
Maverick, Mary, 121, 122, 123, 124, 125
    Samuel A., 16
Maxey, Senator S. B., 123
McCellan's Creek, 77
McGowan, Private, 105
Mckerrow, Isabella, 129
    J. R., 129
McLain (rancher), 80
McLaughlin, Captain N. B., 86, 87 103
McLauren, Mrs., 118, 119
McMartin, (Rancher), 67
McNaught, Captain John S., 116
McRae, Lieutenant C. D., 50,

Medicine Lodge Council, 6
Menger Bar, 131
Merrian, Major Henry C., 97
Merritt, General Wesley, 73, 79
Mescalero Apache Indians, 85
Mescalero Indians, 93
Mexican Army, 93, 115, 116, 117
Mexican Indians, 65
Mexicans, 2, 25, 58, 62, 74, 85, 86, 88, 92, 93, 94, 99
Mexican War, 12
Mexico [throughout book]
Mickle, Hans, 88, 89
Miller, William, 97
Mission San Saba, 13
Missouri, 162
Mizner, Major J. K., 87
Moborg, Private Peter, 130
Mosely, Major General George Van Horn, 157
Moses, Assistant Surgeon Israel, 28
Mounted Rifles, [U.S.], 15, 19, 21, 32
Mud Creek, 65
Mukewarah (Comanche Chief), 6
Mulligan's Bend, 149
Myer, Dr. Albert J., 25, 32, 33, 35, 38

N

Nacimiento, 72
Nance, Bobby, 129
Nance and Struder, 129
Neighbors, Mr., 42
Neill, Colonel Thomas, 102
Nelson, Pop, 150
New Mexico, 42, 52, 78, 118
New York Infantry, 76
Ninth Cavalry, 58, 63, 73, 75, 78, 79, 98, 99
Norris, Assistant Surgeon Basil, 33, 34, 42
Nueces River, 64, 73, 94, 109, 143

O

Oakes, James, 22, 137
O'Brien, Sergeant, 84
Ojinaga, 132
Okeechobee Swamp, 70

Oklahoma Territory, 70, 92
Old Dutch Waterhole, 50, 51
112the Cavalry, 166, 167
118th Infantry, 76
148th Field Artillery, 166
Ord, Brigadier General E.O.C., 108, 113, 114, 115
Osceola, (Seminole Chieftan), 68, 69
Ozona, 145

P

Pah-hah-yuco (Comanche Leader), 5, 6
Paine, Private Adam, 106, 108
Paint Cave, 112
Pair, Private William, 88
Palmito Ranch, 56
Palo Duro Canyon, 105, 106
Panhandle (of Texas), 77, 100, 101
Parker, Quanah, 101
Partisan Rangers, 166
Patton, Bea, 156
    General George S., 152, 153, 154, 155, 156, 157, 158, 159, 160, 161, 171, photo–190
Payne, (Trumpeter) Isaac, 111, 112, 113
    Monroe, 134, 135, 136, 137
    Titus, 108
Payne's Landing, 68
Pearl Harbor, 158, 161, 162, 163
Pecan Springs, 65
Pecos River, 32, 63, 65, 71, 111, 112, 113, 119, 132
Pecos Valley, 58
Pedersen, Private Audeos, 130
Perin, Assistant Surgeon Glover, 28, 29
Perry, Captain Frank W., 72
Pershing, [General] John J. (Blackjack), 155
Phillips, Joseph, 76
Piedras Pintos Creek, 81
Pierson, Major E. P., 143, 144
Plummer, Rachel, 4
Potawatomi Indians, 92

Prairie Rangers, 166
Price, W. E., 16
Prichards, John P., 22, 23

R

Rabago, Philipe de, 13, 14
Ralston, Captain James H., 10
Ramos, Pablo, 56
Rangers, 35
Read, Robert Doddridge, iv
Red River, 4, 57, 100, 104, 106
Reid, Robert, 69
Reiss, Allen, 118, 119
Remington, Frederick, 75, 110
Remolina, [Remolino], 94, 117
Resurreccion (Mexico), 74
Rey Molina, 83
Richardson, R. C., photo–197
Ringgold Barracks, 28, 29, 30
Rio Grande [entire book]
Rio Grande Valley,
Roach and Company, 129
Robenson,, General, photo–199
Roosevelt, President Theodore, 131, 145, 171
Rough Riders, 131
Ruckman, Major General John W., 139
Ruede, Sophia Dorothea, 70

S

Sabinal, 157
Sabinal Vanio, 94
San Antonio, 10, 11, 18, 20, 36, 44, 51, 62, 94, 109, 115, 126, 140, 145, 156
San Antonio–El Paso Road,
San Antonio River, 94, 114, 115
San-da-ve (Lipsan sub-chief), 119
San Diego River, 116
San Felipe, 65
San Miguelito, 156
San Rodriquez River, 117
San Vicente, 133, 135
Sansom, John, 50, 51
Santa Anna, 26
Santa Anita, 137, 138
Santa Rosa, 93
Santa Rosa Mountains, 58, 83, 85
Satanta [Kiowa Chieftain], 100,

Index

101
Schofield, Major General J. M., 125
Schouburgh, Private Wilhelm, 130
Schuhardt, William, 90
Schwandner, Albert, 55, 56
Second Battle of Manassas, 77
Second Cavalry, 22, 37, 47
Second Texas Cavalry, 49
Sells, Lieutenant, 15, 17
Seminole Indians, 68, 69, 70, 72, 109, photo–189, 190
Seminole Mexicans, photo–183
Seminole Negroes, 67, 68, 69, 70, 71, 72, 75, 76,, 81, 83, 84, 85, 87, 95, 96, 97, 98, 103, 106, 107, 108, 109, 110, 113, 114, 115, 117, 118, 120, 135, 164, 171, photo–188
Seminole War, 44
Seventh Cavalry, 78, 163
Shafter, Lieutenant Colonel William R., 94, 113, 114, 115, 116
Shahan, James T. (Happy), 167, 168, 169
Shenandoah Campaign, 77
Sheridan, General Phillip, 57, 77, 79, 80, 82, 89, 93, 98, 106, 122, 124, 125, 171
Sherman, General William T., 77, 78, 93, 123, 125
Shield, Willliam (photo Seminole Negro Scouts), 189
Sibley, Colonel W. W., 135, 136, 137, 138
Sierra Majado, 137, 138
Sierro del Burro Mountains, 119
Sixth Army Corps, 77, 167
Sixth Cavalry, 57, 102, 157
Sixth Infantry, 15
Sixty-ninth Regular Artillery Antiaircraft, 157
Sixty-second Infantry, 56
Slayden, Congressman, 144
Smith, Major H. H., 139, 140
Major, 15
Smithwick, Noah, 6

Smyth, Sergeant Charles, 134
Snake Warrior (Kibbitts), 72
Soldier's Creek, 65
Sonora, 145
Southern [Military] Department, 133
Southern Pacific Railroad, 126, 143, 144
Spofford Junction, 127, 128, 143, 144, 148, 170
Staked Plains, 77, 100, 102, 103, 106
Stanley, General, 119
Steans, General, photo–199
Stiller, Lieutenant, 154
Swift, Surgeon Ebenezer, 36

T

Taft, President William Howard, 133, 145
Tariases, 136
Taylor, General Zachary, 44, 70
Teel, Captain T. T., 48
Ten Bears (Comanche Chief), 6
Terreno Desconocido, 82, 83, 95
Texas Mounted Rifles, 166
Texas Rangers, 1, 2, 7, 11, 12, 32, 132, 133, 169
Third Army, 157, 158, 160
Third Texas Volunteers, 131
Thompson, Ben, 51, 52, 53, 54
    Lieutenant (Hurricane Bill), 102, 104, 105
"Thunderbolt," The, (Bullis), 76
Thurino, Chief [Lipan], 21
Tompkins, D. C., 16
Tonkawa Indians, 12, 13, 71, 102, 103
Torrey's Trading House, 5
Trans-Pecos, 33, 35, 68, 90, 97, 100
Trail of Tears, 70
Travis, Colonel William B., 126
    [his son, Charles Edward], 126
Truman, President Harry S, 162
Tule Canyon, 101, 102, 103
Turby, Private, 28
Twelfth Cavalry, 157, 163, photo–200

Twenty-fifth Cavalry, 71
Twenty-fourth Infantry, 61, 111, 114
Twenty-third Infantry, 130
Twiggs, General David E., 41, 42, 47, 48

U

United States Mounted Rifles, [See Mounted Rifles]
Ute Indians, 7
Uvalde, 55, 67

V

Valdez, General,
Valois, Lieutenant, 74
Van, Green, 80, 86
Veltman, Henry J., 45
Villa de Dolores, 15, 26, 27
Villa, Pancho, 132, 133, 134, 155, 171
Vinson, Lieutenant, 73
Von Rundstedt, 160
Voorhis, Major General Daniel Van, 158

W

Wainwright, General Jonathan, 150, 151, 152, 158, 161, 162, 171, photos–192, 198, 199, 204, 206
    Mrs., photo–191
Walker, Sergenat L. B., photo–210
    Dr. Mary, 129
Walton, Sheriff J. Allen, 133

Ward, Captain D. A., 67, 68
    Sergeant John, 111, 112
Washington, Corporal George, 107
Washo Labo, (Lipan Chief), 114
Watts, Dr., 59
West Point, iv, 17, 72, 77
"Whirlwind, The," 76, 110, 116
Whitfield Legion, 166
Wibarger, Josiah, 3
Wichita Indians, 12, 13
Wilcox, Captain John E., 57, 63
    Captain, 63
Williams, Colonel Thomas G., 92, 93
Wilson, Ben (photo Seminole Negro Scouts), 189
Windus, C. A., 128
Wolf, Jeb, 45
Woman Heart [Kiowa Chief], 102
Wood, C. D., 134
    Picayune John, 95, 96, 99
Worth, General, 70
Wyrick, George, 148, 149, 150, 152, 156

X

Ximinez [Father], 14

Y

Yellow Wolf (Comanche Chieftain), 1
Young, Captain S. B. M., 116

Z

Zaragoza, 85

Editor: Shirley Denise Ratisseau

```
R0127702518 txr        T
                       355
                       .70974
                       7
                       P672
Pirtle, Caleb
The lonely sentinel : Fort
    Clark, on Texas's west
```